ISSN 0044-2267

# ZEITSCHRIFT FÜR ANGEWANDTE MATHEMATIK UND MECHANIK

## INGENIEURWISSENSCHAFTLICHE FORSCHUNGSARBEITEN



BAND 60 1980 HEFT 1

**AUS DEM INHALT**



THEORIA CUM PRAXI

A K A D E M I E - V E R L A G · B E R L I N

ZAMM Bd. 60 Nr. 1 S. 1—64 Berlin, Januar 1980

EVP 18,— M 34115

## INHALT

### Hauptaufsätze



### Kleine Mitteilungen



Wir bitten, **Manuskriptsendungen** zweifach (Original und eine Kopie, sprachlich einwandfrei, Formeln mit Maschine oder in Druckschrift geschrieben) an folgende Anschrift zu richten:

> Zeitschrift für Angewandte Mathematik und Mechanik z. Hd. Herrn Prof. Dr. Günter Schmidt
> Zentralinstitut für Mathematik und Mechanik an der Akademie der Wissenschaften der DDR
> DDR-1080 Berlin, Mohrenstraße 39.

Zu den Arbeiten, die als Hauptaufsätze bestimmt sind, ist auf gesondertem Blatt eine Zusammenfassung von 5 bis 10 Zeilen in englischer und (möglichst) deutscher und russischer Sprache beizufügen. Ausführliche Hinweise für die Autoren, um deren strikte Berücksichtigung gebeten wird, finden sich im Anschluß an das Inhaltsverzeichnis des Jahrganges **59** (1979).

Die Autoren erhalten von den Hauptaufsätzen 75, von den Kleinen Mitteilungen 25 Sonderdrucke ohne Berechnung, darüber hinaus weitere Sonderdrucke gegen Berechnung.

Der Verlag behält sich für alle Beiträge das Recht der Vervielfältigung, Vorbereitung und Übersetzung vor.

Bestellungen sind zu richten

— **in der DDR** an den Postzeitungsvertrieb; an eine Buchhandlung oder an den AKADEMIE-VERLAG, DDR-1080 Berlin, Leipziger Str. 3—4
— **im sozialistischen Ausland** an eine Buchhandlung für fremdsprachige Literatur oder an den zuständigen Postzeitungsvertrieb
— **in der BRD und Westberlin** an eine Buchhandlung oder an die Auslieferungsstelle KUNST UND WISSEN, Erich Bieber, 7000 Stuttgart 1, Wilhelmstraße 4—6
— **in Österreich** an den Globus-Buchvertrieb, 1201 Wien, Höchstädtplatz 3
— **in den übrigen westeuropäischen Ländern** an eine Buchhandlung oder an die Auslieferungsstelle KUNST UND WISSEN, Erich Bieber GmbH, Ch-8008 Zürich/Schweiz, Dufourstraße 51
— **im übrigen Ausland** an den Internationalen Buch- und Zeitschriftenhandel; den Buchexport, Volkseigener Außenhandelsbetrieb der Deutschen Demokratischen Republik, DDR-7010 Leipzig, Postfach 160, oder an den AKADEMIE-VERLAG, DDR-1080 Berlin, Leipziger Straße 3—4

ZEITSCHRIFT FÜR ANGEWANDTE MATHEMATIK UND MECHANIK
Herausgeber und Chefredakteur: Prof. Dr. Günter Schmidt. Redaktion: Dr. Winfried Heinrich, Dr. Horst Weinert, Dipl.-Math. Friedhild Dudel, Helga Rühl, Zentralinstitut für Mathematik und Mechanik der Akademie der Wissenschaften der DDR. Verlag: Akademie-Verlag, DDR-1080 Berlin, Leipziger Straße 3—4; Fernruf: 2236229 oder 2236221 Telex-Nr.: 114420; Bank: Staatsbank der DDR, Berlin, Kto.-Nr.: 6836-26-20712. Anschrift der Redaktion: Zentralinstitut für Mathematik und Mechanik der Akademie der Wissenschaften, DDR-1080 Berlin, Mohrenstraße 39; Fernruf: 2000561.

Veröffentlicht unter der Lizenznummer 1282 des Presseamtes beim Vorsitzenden des Ministerrates der Deutschen Demokratischen Republik. Gesamtherstellung: VEB Druckerei „Thomas Müntzer", 5820 Bad Langensalza. Erscheinungsweise: Die Zeitschrift für Angewandte Mathematik und Mechanik erscheint monatlich. Die 12 Hefte eines Jahres einschließlich Tagungsheft bilden einen Band. Bezugspreis je Band 360,— M zuzüglich Versandspesen (Preis für die DDR 216,— M). Bezugspreis je Heft 30,— M (Preis für die DDR 18,— M). Bestellnummer dieses Heftes: 1009/60/1.



AN (EDV) 35937

ZAMM 60, 1 – 5 (1980)

B. D. CRAVEN

# Strong Vector Minimization and Duality

*In der Terminologie eines Ordnungskegels wird das restringierte Minimum einer vektorwertigen Funktion definiert. Die gewählte Definition führt auf ein vektorielles Analogon des Kuhn-Tucker-Theorems und auf einen Dualitätssatz, in dem das duale Problem eine vektorwertige Zielfunktion besitzt. Ferner wird die schwache Dualität durch geeignete Kegelordnungen definiert. Darüberhinaus werden ein vektorwertiges inverses Dualitätstheorem und ein vektorielles Quasi-Dualitätstheorem bewiesen, bei dem keine Konvexität erforderlich ist. Die Ergebnisse werden in Beziehung gesetzt zu Störungen der Zielfunktion in einem Minimalproblem.*

*A constrained minimum of a vector valued function is defined, in terms of an ordering cone. The definition chosen leads to a vector analog of the Kuhn-Tucker theorem, and to a duality theorem where the dual problem has a vector valued objective function, and weak duality is defined by appropriate cone orderings. Also proved are a vector valued converse duality theorem, and a vector quasiduality theorem which does not require convexity. The results are related to perturbations of the objective function in a minimization problem.*

Определяется условный минимум векторно-значной функции в терминах упорядочения порождающего конусом. Выбранное определение приводит до векторного аналога теоремы Куна-Таккера и до теоремы двойственности, где двойственная задача имеет векторно-значную целевую функцию. Слабая двойственность определяется подходящим упорядочением порождающем конусом. Тоже доказывается векторно-значная обратная теорема двойственности и векторная теорема квази-двойственности в которой выпуклость не требовается. Результаты ставится в связь с возмущением целевой функции в одной минимальной задаче.

## 1. Introduction

Consider the vector-valued minimization problem:

$$\text{Minimize } f(x) \text{ subject to } x \in H\,, \tag{1}$$

where $H \subset \mathbb{R}^n$, $f : \mathbb{R}^n \to \mathbb{R}^r$ is a continuous function, with $r > 1$, and $P$ is a closed convex cone in $\mathbb{R}^r$ (in particular $P$ may be the nonnegative orthant $\mathbb{R}^r_+$ in $\mathbb{R}^r$). A global minimum of (1) at $x = \bar{x} \in H$ has been defined in several ways, which may be expressed as follows. The problem (1) has a *vector minimum* at $\bar{x}$ (see [11, page 216], [12], [13]) if there exists no $x \in H$ for which $f(\bar{x}) - f(x) \in P \setminus \{0\}$. The problem has a *weak minimum* at $\bar{x} = x$ (see [2], [1, page 33]) if there exists no $x \in H$ for which $f(\bar{x}) - f(x) \in \operatorname{int} P$, where $\operatorname{int} P$ denotes the interior of $P$. When, as in [8], [9], [13],

$$f(x) - f(\bar{x}) \in P \quad \text{for all } x \in H\,, \tag{2}$$

the point $\bar{x}$ will be called a *strong minimum* of (1), to distinguish this case from the other two. These are *global* minima; definitions for the corresponding *local* minima are obtained from those above by replacing $H$ by $H \cap N$, where $N$ is a sufficiently small neighbourhood of $\bar{x}$. Clearly, a strong minimum implies a vector minimum, and a vector minimum implies a weak minimum. Necessary conditions for a weak minimum are known [1], [2], [4], [5], in terms of a real-valued LAGRANGIAN.

In the present paper, a strong minimum is described, using a vector-valued LAGRANGIAN; this leads to duality and converse duality theorems, where the dual problem also has a vector-valued objective function. Consider the minimization problem:

$$\text{Minimize } f(x) \text{ subject to } h(x) \in T\,, \tag{3}$$

where $f : \mathbb{R}^n \to \mathbb{R}^r$ and $h : \mathbb{R}^n \to \mathbb{R}^m$ are differentiable functions; $P \subset \mathbb{R}^r$ and $T \subset \mathbb{R}^m$ are closed convex cones; and a (local) strong minimum is defined, as above, using the cone $P$, with $H = \{x : h(x) \in T\}$. The domain of $f$ and $h$ may be reduced to an open convex subset of $\mathbb{R}^n$, without changing the results.

When $P = \mathbb{R}^r_+$, strong minimization with respect to $P$ means simultaneous minimization of each component of $f(x)$. This requirement is too stringent for most applications of vector-valued minimization. More relevant cases are obtained by choosing $P$ as a suitable polyhedral cone. Let $P \subset \mathbb{R}^r$ be the polyhedral cone defined by $y \in P$ iff $Cy \in \mathbb{R}^s_+$, where $C$ is a given matrix, not necessarily square or invertible. Assume the constraint $h(x) \in T$ to be $h_i(x) \geqq 0$ $(i = 1, 2, \dots, m)$. Then a strong minimum of $f(x)$ with respect to $P$ is equivalent to a strong minimum of $Cf(x)$ with respect to $\mathbb{R}_s^+$. For the latter problem, the vector KUHN-TUCKER conditions (Theorem 8) for a strong minimum at $x = \bar{x}$ lead to

$$Cf'(x) = Wh'(x)\,, \qquad Wh(x) = 0\,, \qquad W \geqq 0\,, \qquad h(x) \geqq 0\,, \tag{4}$$

where $W \geqq 0$ means that each component of the matrix $W$ is $\geqq 0$. *If $C$ happens to be invertible, then (4) leads to the vector* KUHN-TUCKER *conditions for the original problem, thus, for a suitable matrix $\bar{V} \in \mathbb{R}^{r \times m}$,*

$$f'(\bar{x}) = \bar{V}h'(\bar{x})\,, \qquad \bar{V}h(\bar{x}) = 0\,, \qquad C\bar{V} \geqq 0\,, \qquad h(\bar{x}) \geqq 0\,. \tag{5}$$

In fact from (Theorem 8), (5) holds without this restriction on $C$. Moreover, *if the problem is also convex, then (5) is also sufficient for a strong minimum.*

Consider, in particular, a linear program with $h(x) = Ax - b$ and

$$f(x) = \begin{bmatrix} c_1^\mathsf{T} x \\ c_2^\mathsf{T} x \end{bmatrix}, \qquad C = \begin{bmatrix} 1 & 0 \\ \beta & 1 \\ \gamma & \delta \end{bmatrix},$$

where $\beta, \gamma, \delta$ are real constants with $\gamma > 0$. Assume that $c_1^\mathsf{T} x$ is minimized, subject to the constraints, at $x = \bar{x}$. Except in some limiting cases, a perturbation of $c_1^\mathsf{T} x$ to $(c_1 + \beta^{-1} c_2)^\mathsf{T} x$ or to $(c_1 + \gamma^{-1} \delta c_2)^\mathsf{T} x$ leaves the optimal basis unchanged, while $\beta$ and $\gamma^{-1}\delta$ remain in suitable intervals. For such $\beta, \gamma, \delta$, the vector function $f(x)$ reaches a strong minimum with respect to the polyhedral cone $C^{-1}(\mathbb{R}^3_+)$.

Let $\mathbb{F}$ denote the space $\mathbb{R}^r$, with the partial ordering $\leqq$ defined by $y \leqq x$ iff $x - y \in P$. Results of ELSTER and NEHSE [8], [9], [10] show that *the following statements are equivalent.*

(a) *$\mathbb{F}$ is order-complete (or conditionally complete).*

(b) *The* HAHN-BANACH *extension property always holds, for linear mappings into $\mathbb{F}$.*

(c) *The* FARKAS-MINKOWSKI *theorem holds with linear functional replaced by linear mapping into $\mathbb{F}$* (see Theorem 0, below, for relevant details).

(d) *a vector-valued* KUHN-TUCKER *saddlepoint theorem holds for a strong minimum of* (3), *assuming that* (3) *is a convex problem, and assuming a constraint qualification; the* LAGRANGIAN *is then $f(x) - Vh(x)$, where the* LAGRANGE *multiplier $V$ is a linear map, which takes $T$ into $P$.*

If $P = \mathbb{R}^r_+$, then $\mathbb{F}$ is order-complete, and these results can be applied. However, there seems to be no general criterion for the order-completeness of $\mathbb{F}$, directly in terms of the cone $P$. For example, it is not clear whether $\mathbb{F}$ is order-complete when $P$ is an arbitrary polyhedral cone. Moreover, the KUHN-TUCKER conditions may hold without order-completeness of $\mathbb{F}$, if other parts of the problem are suitably restricted. Also duality and related theorems may hold, as shown below, when (unlike [8], [9]) different ordering cones are used for the given (primal) problem and its vector dual.

## 2. Preliminary results and definitions

Vectors $x \in \mathbb{R}^n$ will be considered as column vectors, and dual vectors as rows. The (FRÉCHET) derivative $f'(x)$ of $f$ is considered as an $r \times n$ matrix of partial derivatives $\partial f_i / \partial x_j$. The *dual cone* of $T$ is the convex cone $T^*$ of all dual vectors $z$ satisfying $zt \geqq 0$ for all $t \in T$. Denote by $M(P, T)$ the set of all linear maps from $\mathbb{R}^m$ into $\mathbb{R}^n$ which map $T$ into $P$. A subset $K \subset \mathbb{R}^r$ will be called a *basic cone* if $K$ is a convex cone whose generators form a basis of $\mathbb{R}^r$. If $K \subset \mathbb{R}^r$ and $Q \subset \mathbb{R}^r$, denote $K + Q = \{k + q : k \in K, q \in Q\}$. The function $h$ is *$T$-convex* if, whenever $0 < \alpha < 1$ and $x, x'$ are in the domain of $h$.

$$\alpha h(x) + (1 - \alpha)\, h(x') - h(\alpha x + (1 - \alpha)\, x') \in T\,. \tag{6}$$

Theorem 0 (Generalized FARKAS Theorem): *Let $A : \mathbb{R}^h \to \mathbb{R}^m$ and $B : \mathbb{R}^h \to \mathbb{R}^r$ be linear maps; let $T \subset \mathbb{R}^m$ and $K \subset \mathbb{R}^r$ be closed convex cones. Assume that either* (i) *$A(\mathbb{R}^h) = \mathbb{R}^m$, or* (ii) *$K$ is a basic cone and $A^\mathsf{T}(T^*)$ is closed. Then*

$$[Az \in T \Rightarrow Bz \in K] \Leftrightarrow [(\exists V \in M(T, K))\; B = VA]\,. \tag{7}$$

Proof: Case (i) follows from [2], Theorem 2.2; and case (ii) from [6], Theorem 11.

Since $\mathbb{R}^n$, ordered by the basic cone $K$, is order-complete, the result of [9] may also be used to prove case (ii). In [9], a different hypothesis to closedness of $A^\mathsf{T}(T^*)$ is assumed; however, it is equivalent, using [6], Theorem 7. It remains an open question whether the hypothesis, in (ii), that $K$ is basic, may be omitted. If it can, then various subsequent results can be simplified. The related results of RITTER [13], applied to the present case, would weaken the conclusion $V \in M(T, K)$ to $V \in M(T \cap \operatorname{ran}(A), K)$, where ran denotes range.

The system $h(x) \in T$ is *locally solvable* at $\bar{x}$ (see [4], [5]) if $h(\bar{x}) \in T$ and, for some $\delta > 0$, whenever the direction $d$ satisfies $h(\bar{x}) + h'(\bar{x})\, d \in T$ and $||d|| < \delta$, there exists a solution $x = \bar{x} + \alpha d + o(\alpha)$ to $h(x) \in T$, valid for all sufficiently small $\alpha > 0$. (Here $\alpha^{-1} o(\alpha) \to 0$ as $\alpha \downarrow 0$.) For this it suffices [4], [14] if $h(\bar{x}) + h'(\bar{x})\, (\mathbb{R}^n) - T$ contains a ball with centre 0. For scalar-valued constraints

$$h_i(x) = 0 \quad (i = 1, 2, \dots, s); \qquad h_i(x) \geqq 0 \quad (i = s + 1, \dots, m)\,, \tag{8}$$

local solvability reduces [5] to the KUHN-TUCKER constraint qualification.

A *strong maximum* of $f(x)$, with respect to the cone $P$, is defined by minimizing $-f(x)$, with respect to $P$. The problem (1) has a *strong quasimin* at $x = \bar{x}$ if $\bar{x} \in H$, and, for some function $\theta : \mathbb{R}^n \to \mathbb{R}^r$ satisfying $\theta(x) = o(||x - \bar{x}||)$,

$$f(x) - f(\bar{x}) - \theta(x) \in P \quad \text{as} \quad x \to \bar{x} \quad \text{with} \quad x \in H\,. \tag{9}$$

Similarly, $f(x)$ has a *strong quasimax* at $x = \bar{x}$, subject to $x \in H$, if $-f(x)$ has a strong quasimin, subject to $x \in H$ (with respect to the same cone $P$). Note that a strong quasimin is defined locally, not globally.

A *quasimin* was defined in [4], as a generalization of local weak minimum; a strong quasimin bears a similar relation to strong local minimum. The relation of quasimin to local minimum for a constrained problem is approximately that of stationary point to local minimum for an unconstrained problem; see [4], Theorem 2.

The symbol $V \geqq 0$, applied to a matrix $V$, means that each component of $V$ is nonnegative.

### 3. Vector duality

Consider the problem:

$$\underset{x \in R^n}{\text{Minimize}}\ f(x) \text{ subject to } h(x) \in T, \text{ (strongly with respect to } P), \tag{10}$$

in relation to the problem:

$$\underset{u \in R^n,\ v \in R^{r \times m}}{\text{Maximize}}\ f(u) - Vh(u) \text{ subject to } f'(u) = Vh'(u),\ V \in M(T, K), \text{ (strongly with respect to } K + Q). \tag{11}$$

Here $K$ is a convex cone containing $P$, as specified below in Theorem 1, and $Q$ is another given convex cone in $R^r$. The cone $K + Q$ is convex, but not necessarily closed.

Theorem 1: *Let $f : R^n \to R^r$ and $h : R^n \to R^m$ be differentiable functions; let $T \subset R^m$ and $P \subset R^r$ be closed convex cones. Let* (10) *reach a local strong minimum at $x = \bar{x}$. Either* (i) *let $h'(\bar{x})$ be surjective (and then let $K = P$), or* (ii) *let the convex cone $h'(\bar{x})^{\mathsf{T}}(T^*)$ be closed, and let $K \supset P$ be a basic cone. Let $h(x) \in T$ be locally solvable at $\bar{x}$. Then, for some $\bar{V} \in M(T, K)$,*

$$f'(\bar{x}) = \bar{V}h'(\bar{x}) \quad \text{and} \quad \bar{V}h(\bar{x}) = 0\,. \tag{12}$$

Proof: Denote $A = [h'(\bar{x}), h(\bar{x})]$ and $B = [f'(\bar{x}), f(\bar{x})]$. Let $z = (d, \beta) \in R^n \times R$ satisfy $Az \in T$. For $\gamma > 0$, $h(\bar{x})\,\gamma + [h'(\bar{x})\,d + h(\bar{x})\,\beta] \in T$. For $\gamma$ large enough, $d' = (\gamma + \beta)^{-1}\,d$ satisfies $\|d'\| < \delta$ and $h(\bar{x}) + h'(x)\,d' \in T$. By local solvability, $h(x) \in T$ has a solution $x = \bar{x} + \alpha d' + o(\alpha)$, for sufficiently small $\alpha > 0$. Since $\bar{x}$ minimizes (10), $f(\bar{x} + \alpha d' + o(\alpha)) - f(\bar{x}) \in P$ for $\alpha + 0$. Since $P$ is a convex cone, $f'(\bar{x})d' + o(\alpha)/\alpha \in P$; since $P$ is closed, $f'(\bar{x})d = (\beta + \gamma)\,f'(x)\,d' \in P$. Thus

$$Az \in T \Rightarrow Bz = f'(x)\,d \in P \subset K\,. \tag{13}$$

From Theorem 0, under either hypothesis (i) or (ii), there exists $\bar{V} \in M(T, K)$ such that $B = \bar{V}A$. Hence (12) follows.

This result is a vector version of the KUHN-TUCKER theorem. The basic cone $K$ is introduced, so that the case (ii) of Theorem 0 can be applied. The next theorem gives a vector analog of weak duality.

Theorem 2: *In problem* (10), *let $f$ be $Q$-convex, where $Q$ is a convex cone in $R^r$; and let $-h$ be $T$-convex. Let $x$ be any feasible point for* (10) *and let $(u, V)$ be any feasible point for* (11). *Then*

$$f(x) - [f(u) - Vh(u)] \in K + Q\,. \tag{14}$$

Proof: Since $f$ is $Q$-convex,

$$\begin{aligned} f(x) - [f(u) - Vh(u)] &= f'(u)\,(x - u) + q - Vh(u) && \text{for some } q \in Q \\ &= Vh'(u)\,(x - u) - Vh(u) + q && \text{by the constraint of (11)} \\ &= k + Vh(x) + q && \text{for some } k \in K \qquad (15) \\ &\in K + Q && \text{since } V \in M(T, K) \text{ and } h(x) \in T, \text{ so } Vh(x) \in K. \end{aligned}$$

To validate the step (15), $h$ is $T$-convex and $V \in M(T, K)$, hence $Vh$ is $K$-convex, hence

$Vh(x) - Vh(u) - Vh'(u)\,(x - u) \in K$ (e.g. [5], Theorem 2.4.2).

The following vector duality theorem then follows.

Theorem 3: *Assume all hypotheses of Theorems 1 and 2. Let* (10) *reach a strong minimum at $x = \bar{x}$. Then* (11) *reaches a strong maximum at $(u, V) = (\bar{x}, \bar{V})$, for some $\bar{V} \in M(T, K)$; the objective functions of* (10) *and* (11) *are equal at these points; and vector weak duality* (14) *holds.*

Proof: From Theorem 1, $(\bar{x}, \bar{V})$ is fesaible for (11), for some $\bar{V} \in M(T, K)$, where $\bar{V}h(\bar{x}) = 0$. From this and vector weak duality (from Theorem 2), $f(\bar{x}) = f(\bar{x}) - \bar{V}h(\bar{x})$, and whenever $(u, V)$ is feasible for (11),

$f(\bar{x}) - [f(u) - Vh(u)] \in K + Q$.

Theorem 4: *Let $f$ be $Q$-convex, and let $-h$ be $T$-convex. Let* (12) *hold, with also $h(\bar{x}) \in T$. Then $\bar{x}$ is a strong minimum of* (10) *with respect to the cone $Q + \bar{V}(T)$.*

Proof: Define $L = f - \bar{V}h$; then $L$ is $(Q + \bar{V}(T))$-convex, by the hypotheses. Whenever $h(x) \in T$, $L(x) - L(\bar{x}) = L'(\bar{x})\,(x - \bar{x}) + n = 0 + n$ for some $r \in Q + \bar{V}(T)$, using (12). Since $\bar{V}h(\bar{x}) = 0$,

$$f(x) - f(\bar{x}) = [L(x) - L(\bar{x})] + \bar{V}h(x) \in (Q + \bar{V}(T)) + \bar{V}(T) = Q + \bar{V}(T)\,.$$

The duality theory of [8], [9] corresponds to the particular case when $P = Q = K + Q$.

### 4. Vector quasiduality

The next theorem omits the convexity requirements on the functions, and replaces strong minimum by strong quasimin.

Theorem 5: *Let $f : R^n \to R^r$ and $h : R^n \to R^m$ be differentiable functions; let $T \subset R^m$ and $P \subset R^r$ be closed convex cones. If* (10) *reaches a strong quasimin, with respect to $P$, at $x = \bar{x}$, and if also $h(x) \in T$ is locally solvable at $\bar{x}$,* *and either hypothesis* (i) *or* (ii) *of Theorem* 1 *holds, then* (12) *holds for some* $\bar{V} \in M(T, K)$. *Conversely, if* $K$ *is a convex cone such that* (12) *holds, for some* $\bar{V} \in M(T, K)$, *then* (10) *reaches a strong quasimin at* $\bar{x}$, *with respect to* $K$.

Remark: If (12) holds with $\bar{V} \in M(T, P)$, then (10) reaches a strong quasimin at $\bar{x}$ with respect to $P$. Theorem 5 generalizes [5], Theorem 5.1.1.

Proof: The proof that a strong quasimin, with side hypotheses, implies (12), differs from the proof of Theorem 1, only by replacing $f(\bar{x} + \alpha d' + o(\alpha)) - f(\bar{x}) \in P$ by $f(\bar{x} + \alpha d' + o(\alpha)) - f(\bar{x}) - \theta(\alpha) \in P$ for $\alpha \downarrow 0$, for some function $\theta$ satisfying $\theta(\alpha) = o(\alpha)$ as $\alpha \downarrow 0$. The remainder of the proof is unchanged.

Conversely, assume (12), but suppose that $\bar{x}$ is *not* a strong quasimin of (10) with respect to $P$. Then, for some sequence $\{z_i\} \subset \mathbb{R}^n$ for which $\{||z_i||\} \to 0$ and $\{h(\bar{x} + z_i)\} \subset T$, whenever $\theta(z) = o(||z||)$,

$$f(\bar{x} + z_i) - f(\bar{x}) - \theta(z_i) \notin K . \tag{16}$$

Now, using (12),

$$\bar{V}h'(\bar{x})\, z_i = \bar{V}[h(\bar{x} + z_i) - h(\bar{x}) + o(||z_i||)] = \bar{V}h(\bar{x} + z_i) - 0 + o(||z_i||) .$$

Hence, using (12) again,

$$f(\bar{x} + z_i) - f(\bar{x}) = f'(\bar{x})\, z_i + o(||z_i||) = \bar{V}h'(\bar{x})\, z_i + o(||z_i||) = v_i + \bar{\theta}(z_i) ,$$

where $v_i = \bar{V}h(\bar{x} + z_i) \in K$, using $\bar{V} \in M(T, K)$, and $\bar{\theta}(z_i) = o(||z_i||)$. Choosing $\theta = \bar{\theta}$ gives

$$f(\bar{x} + z_i) - f(\bar{x}) - \theta(z_i) \in K \ (i \to \infty) . \tag{17}$$

Since (17) contradicts (16), it follows that $\bar{x}$ is a strong quasimin of (10).

Adapting the definition of quasidual in [4], problem (11) will be called a *vector quasidual* of (10) if, whenever (10) reaches a strong quasimin *at* $\bar{x}$, (11) reaches a strong quasimax at some $(\bar{u}, \bar{V})$, and the two objective functions are equal.

Theorem 6: *Assume the hypotheses of the first part of Theorem* 5. *Then problem* (11) *is a vector quasidual of problem* (10).

Proof: Let (10) reach a strong quasimin at $\bar{x}$. Given the hypotheses, (12) holds, from Theorem 5, with some $\bar{V}$. Let $(u, V)$ be feasible for (11), with $u = \bar{x} + p$ and $v = \bar{V} + w$. Then, since $\bar{V}h(\bar{x}) = 0$,

$$f(x) - [f(u) - Vh(u)] = -f'(x)\,p - o(||p||) + Wh(\bar{x}) + \bar{V}h'(\bar{x})\,p + o(||p|| + ||W||) = k + o(||p|| + ||W||),$$

for some $k \in K$, by (12).

Hence (11) has a strong quasimax at $(x, V)$. Since also $f(\bar{x}) = f(\bar{x}) - \bar{V}h(\bar{x})$, (11) is a strong quasidual of (10).

## 5. Converse duality

The following theorem generalizes the converse duality theorems of [3], and [5] Theorem 4.8.1, to strong minima and maxima.

Theorem 7: *Let* $P, Q \subset \mathbb{R}^r$ *and* $T \subset \mathbb{R}^m$ *be closed convex cones. Let* $f : \mathbb{R}^n \to \mathbb{R}^r$ *and* $-h : \mathbb{R}^n \to \mathbb{R}^m$ *be differentiable functions, respectively Q-convex and T-convex. Let* (11) *reach a strong maximum, with respect to* $P$, *at* $(u, V) = (\bar{x}, \bar{V})$, *where also* $Q + \bar{V}(T) \subset P$, *and* $P \cap (-P) = \{0\}$. *Whenever* $\bar{V} + W \in M(T, P)$ *and* $\alpha$ *is sufficiently small positive, assume that the equation* $f'(\bar{x} + \xi(\alpha)) = (\bar{V} + \alpha W)\, h'(\bar{x} + \xi(\alpha))$ *has a solution* $\xi(\alpha)$, *such that* $\xi(\alpha) \to 0$ *as* $\alpha \downarrow 0$. *Then* (10) *reaches a strong minimum, with respect to* $P$, *at* $x = \bar{x}$.

Proof: Let $\bar{V} + W \in M(T, P)$, and let $0 < \alpha < 1$. Since $M(T, P)$ is a convex cone, and $\bar{V} \in M(T, P)$,

$$\bar{V} + \alpha W = (1 - \alpha)\, \bar{V} + \alpha(\bar{V} + W) \in M(T, P) + M(T, P) = M(T, P) .$$

For sufficiently small $\alpha$, since $(\bar{x}, \bar{V})$ is a strong maximum of (11) with respect to $P$,

$$O = [f(\bar{x} + \xi(\alpha)) - (\bar{V} + \alpha W)\, h(\bar{x} + \xi(\alpha))] - [f(\bar{x}) - \bar{V}h(\bar{x})] + p \quad \text{for some } p \in P ,$$
$$= f'(\bar{x})\, \xi(\alpha) + q - \bar{V}h'(\bar{x})\, \xi(\alpha) + r + p - \alpha Wh(\bar{x} + \xi(\alpha))$$

for some $r \in \bar{V}(T)$ and some $q \in Q$, using convexity. Hence

$$O = 0 - \alpha Wh(\bar{x} + \xi(\alpha)) + s ,$$

where $s = p + q + r \in P + Q + \bar{V}(T) \subset P$, by hypothesis. Hence $Wh(\bar{x} + \xi(\alpha)) \in P$. Since $P$ is closed, and $\xi(\alpha) \to 0$, $Wh(\bar{x}) \in P$. This holds, in particular, for $W \in M(T, P)$.

Suppose that $h(\bar{x}) \notin T$. Since $T$ is a closed convex cone, there exists $\varphi \in T^*$ such that $\varphi(h(\bar{x})) < 0$ (e.g. [5], Theorem 2.2.4). Define $\theta \in M(T, P)$ by $(\forall t \in T)\ \theta(t) = \varphi(t)\, p_0$, where $p_0$ is a fixed nonzero element of $P$. Then, for some $\beta > 0$, $\theta(h(\bar{x})) = -\beta p_0 \notin P$ since $P \cap (-P) = \{0\}$. Since this contradicts $(\forall W \in M(T, P))\ Wh(\bar{x}) \in P$, it follows that $h(\bar{x}) \in T$.

Setting $W = -\frac{1}{2}\bar{V}$, $\bar{V} + W \in M(T, P)$, so $-\frac{1}{2}\bar{V}h(\bar{x}) \in P$. But also $\bar{V}h(\bar{x}) \in P$. Hence $\bar{V}h(\bar{x}) \in P \cap (-P) = \{0\}$. Therefore (12) holds, with $h(\bar{x}) \in T$. From Theorem 4, it follows that $\bar{x}$ is a strong minimum of (10) with respect to $P$.

## 6. Application with a polyhedral cone

In this section, assume the constraint $h(x) \in T$ in the form

$$h_i(x) \geqq 0 \qquad (i = 1, 2, \dots, m) . \tag{18}$$

Theorem 8: *Let $P \subset \mathbb{R}^r$ be the polyhedral cone defined by $y \in P$ iff $Cy \in \mathbb{R}^s_+$, where $C$ is a given matrix, not necessarily square or invertible. Let $f : \mathbb{R}^n \to \mathbb{R}^r$ satisfy either* (a) *$f$ is linear, or* (b) *$f$ is differentiable and each component of $Cf$ is convex, or* (c) *$f$ is twice differentiable and $Cf''(x)$ is positive semidefinite for each $x$. Let $x = \bar{x}$ satisfy* (18); *let each $-h_i$ be convex; let the gradients $h_i'(\bar{x})$ be linearly independent for the constraints active at $\bar{x}$ (thus when $h_i(\bar{x}) = 0$). Then $f(x)$ reaches a strong minimum at $x = \bar{x}$, subject to the constraints* (18), *with respect to $P = C^{-1}(\mathbb{R}^s_+)$, if and only if there exists a matrix $\bar{V} \in \mathbb{R}^{r \times m}$ satisfying*

$$f'(\bar{x}) = \bar{V}h'(\bar{x}) , \qquad \bar{V}h(\bar{x}) = 0 , \qquad C\bar{V} \geqq 0 , \qquad h(\bar{x}) \geqq 0 . \tag{19}$$

Proof: Under any of the hypotheses (a), (b), or (c), $f$ is $P$-convex. (For (c) see e.g. [5], page 29.) The inactive constraints from (18) do not affect a strong minimum at $\bar{x}$; hence Theorem 1 may be applied with the inactive constraints omitted. Case (i) applies, by the linear independence hypothesis. If $\bar{x}$ is a strong minimum, then Theorem 1 shows that (12) holds, for some matrix $\bar{V} \in M(\mathbb{R}^+_n, P)$. The latter holds, when $P = C^{-1}(\mathbb{R}^s_+)$, if and only if $C\bar{V}$ maps each basis vector $e_i$ of $\mathbb{R}^m_+$ into $\mathbb{R}^s_+$, thus if and only if $C\bar{V} \geqq 0$. Thus (19) holds. Conversely, if (19) holds, then $\bar{V} \in M(\mathbb{R}^m_+, P)$, also $f$ is $P$-convex, and Theorem 4 shows that $\bar{x}$ is a strong minimum with respect to $P + \bar{V}(\mathbb{R}^m_+) = P$.

If $h(x) \in T$ includes also equality constraints, then, as well as $C\bar{V} \geqq 0$, each column $\bar{V}_j$ of $\bar{V}$ which corresponds to an equality constraint must satisfy $C\bar{V}_j = 0$.

Consider now a convex nonlinear program, as in Theorem 7. Suppose that (19) holds for a particular matrix $\bar{V}$, say $\bar{V}_0$. Then $\bar{x}$ minimizes each component of $\bar{V}_0 f(x)$. Now let $\bar{V}$ be any other matrix satisfying (19). It follows that $\bar{x}$ also minimizes $\bar{V}f(x)$. Thus Theorem 8 specifies a set of matrices $C$, considered as perturbing the objective function, which leave the optimum unchanged. This applies in particular to linear programming, as discussed above in section 1.

### Acknowledgement

I am indebted to a referee for bringing references [8], [9], [10] and [13] to my attention, and for other comments and suggestions.

### References

1 BORWEIN, J. M., Optimization with respect to partial orderings, D. Phil. thesis, Oxford University 1974.
2 CRAVEN, B. D., Nonlinear programming in locally convex spaces, J. Optim. Theor. Appl. 10, 197—210 (1972).
3 CRAVEN, B. D., Converse duality in BANACH space, J. Optim. Theor. Appl. 17, 229—238 (1975).
4 CRAVEN, B. D., LAGRANGEan conditions and quasiduality, Bull. Austral. Math. Soc. 16, 325—339 (1977).
5 CRAVEN, B. D., Mathematical Programming and Control Theory, Chapman and Hall, London 1978.
6 CRAVEN, B. D.; KOLIHA, J. J., Generalizations of FARKAS' theorem, SIAM. J. Math. Analysis 8, 983—997 (1977).
7 CRAVEN, B. D.; MOND, B., On converse duality in nonlinear programming, Operations Res. 19, 1075—1078 (1971).
8 ELSTER, K.-H.; NEHSE, R., Nichtkonvexe Optimierungsprobleme, 21. Intern. wiss. Koll. TH Ilmenau 1976, Sektion Mathematik, Rechentechnik und ökonomische Kybernetik, pp. 69—72.
9 ELSTER, K.-H.; NEHSE, R., Necessary and sufficient conditions for the order-completeness of partially ordered vector spaces, Math. Nachr. 81, 301—311 (1978).
10 NEHSE, R., The HAHN-BANACH property and equivalent conditions, Comment. Math. Univ. Carolinae 19, 165—177 (1978).
11 KARLIN, S., Mathematical Methods and Theory in Games, Programming and Economics, Vol. 1, Addison-Wesley, Reading, Mass 1959 (Page 216).
12 MANGASARIAN, O. L.; SUTHERLAND, W. R. S., Solution of the linear inverse vector optimization problem by a single linear program, Math. Programming 15, 232—235 (1978).
13 RITTER, K., Optimization theory in linear spaces. Part III: Mathematical programming in partially ordered BANACH spaces, Math. Ann. 184, 133—154 (1970).
14 ROBINSON, S. M., Stability theory for systems of inequalities. Part II: Differentiable nonlinear systems, SIAM. J. Numer. Analysis 13, 497—513 (1976).

Eingereicht: 1. 12. 1978, revidierte Fassung 25. 6. 1979

*Anschrift*: Professor Dr. B. D. CRAVEN, University of Melbourne, Department of Mathematics, Parkville, Victoria 3052, Australien

ZAMM 60, 7 – 17 (1980)

F. VOGL

# Über ein System linearer Funktional-Differentialgleichungen

*In der Arbeit werden Systeme linearer Funktional-Differentialgleichungen der Form* $\boldsymbol{y}'(x) = A\boldsymbol{y}(\lambda x)$, $0 < \lambda < 1$, *untersucht. Die Elemente* $a_{jk}$ *der Matrix A sind komplexe Konstante, während die Variable x reell oder komplex sein kann. Vorerst wird die Struktur aller ganzen Lösungen angegeben. Anschließend wird für* $x \neq 0$ *ein System spezieller analytischer Lösungen konstruiert, mit dessen Hilfe alle weiteren Lösungen dargestellt werden können. Außerdem werden die asymptotischen Eigenschaften (für* $x \to \infty$*) aller Lösungen diskutiert, und es wird gezeigt, daß jeweils genau eine Lösung mit einem bestimmten, geeignet vorgegebenen asymptotischen Verhalten existiert.*

*The paper discusses systems of linear functional-differential equations of the form* $\boldsymbol{y}'(x) = A\boldsymbol{y}(\lambda x)$, $0 < \lambda < 1$, *where the elements* $a_{jk}$ *of the matrix A are complex constants. The variable x may be real or complex. At first the form of all entire solutions is given. Subsequently for* $x \neq 0$ *a system of special analytic solutions is constructed by means of which all further solutions may be represented. Finally the asymptotic properties (if* $x \to \infty$*) of all solutions of the system are investigated. Furthermore it is shown that, given a specific asymptotic behavior, there is one and only one solution which possesses that asymptotic behavior.*

В работе рассматриваются системы линейных функционально-дифференциальных уравнений вида $\boldsymbol{y}'(x) = A\boldsymbol{y}(\lambda x)$, $0 < \lambda < 1$. Здесь элементы матрицы $A$ комплексные постоянные, а переменная $x$ может быть либо комплексной, либо вещественной. Сначала описывается структура всех целых решений. После этого, для случая $x \neq 0$, конструируется система специальных аналитических решений, с помощью которой можно представить все остальные решения. Кроме этого, исследуются асимптотические свойства (при $x \to \infty$) всех решений и показывается, что существует единственное решение, имеющее надлежащим образом предписываемое асимптотическое поведение.

## TEIL I: ANALYTISCHE LÖSUNGEN

### Einleitung

Von CARR und DYSON [2] bzw. ENG BIN LIM [3] wurde das asymptotische Verhalten ($x \to \infty$) der Lösungen von Systemen linearer Funktional-Differentialgleichungen der Form

$$\boldsymbol{y}'(x) = A\boldsymbol{y}(\lambda x) + B\boldsymbol{y}(x)\,, \qquad 0 < \lambda < 1\,, \quad x \geqq 0\,, \quad a_{jk}, b_{jk} \in \mathbb{C}\,, \tag{1}$$

unter der Voraussetzung nicht verschwindender Eigenwerte der Matrix $B$ untersucht. Dabei wurden ähnliche Resultate wie für die in [5], [6] behandelte skalare Gleichung

$$y'(x) = ay(\lambda x) + by(x)\,, \qquad x \geqq 0\,, \tag{2}$$

bewiesen. Sind $A$ und $B$ reelle konstante Matrizen, so ist System (1) ein mathematisches Modell für das dynamische Verhalten von Stromabnahmebügel und Oberleitung bei elektrischen Bahnen (vgl. [7]).

Für eine vollständige Analyse des Systems (1), d. h. unter Einschluß des Falles verschwindender Eigenwerte von $B$, ist eine Kenntnis der Verhältnisse bei Systemen der Form $\boldsymbol{y}'(x) = A\boldsymbol{y}(\lambda x)$ erforderlich. Es ist bemerkenswert, daß die Diskussion dieses Systems einen größeren mathematischen Aufwand erfordert, als die des Systems (1) für nicht verschwindende Eigenwerte von $B$.

Das Ziel dieser Arbeit ist das Studium der folgenden beiden Anfangswertprobleme (AWP), wobei vor allem das asymptotische Verhalten der Lösungen für $x \to \infty$ im Mittelpunkt des Interesses steht.

$$\boldsymbol{y}'(x) = A\boldsymbol{y}(\lambda x)\,, \qquad 0 < \lambda < 1\,, \tag{3}$$

$$\boldsymbol{y}(0) = \boldsymbol{y}_0 \in \mathbb{C}^r \tag{4}$$

Die Systemmatrix $A$ wird als konstante $r \times r$-Matrix mit Elementen $a_{jk} \in \mathbb{C}$ vorausgesetzt, während die Variable $x$ reell oder komplex sein kann. Die *Lösungen* dieses AWP sind für komplexes $x$ ganze (Vektor-)Funktionen der Ordnung Null, deren Einschränkung auf die reelle Achse die Lösungen für reelle Argumente liefert. (Vgl. **1**). Durch geeignete Definition des Lösungsbegriffes kann neben (3), (4) für $x \in \mathbb{R}_+$ noch folgendes AWP betrachtet werden:

$$\boldsymbol{y}'(x) = A\boldsymbol{y}(\lambda x)\,, \tag{3}$$

$$\boldsymbol{y}(x)|_{I_0} = \boldsymbol{\alpha}(x) \in C(I_0)\,, \qquad I_0 = [\lambda x_0, x_0]\,, \quad x_0 \in \mathbb{R}_+\,, \quad C(I_0)\text{: Raum der stetigen Funktionen auf } I_0. \tag{5}$$

Als *Lösung* des AWP (3), (5) wird jede komplexwertige Funktion $\boldsymbol{y}(x)$ mit folgenden Eigenschaften bezeichnet:

$$\boldsymbol{y}(x) \in C([\lambda x_0, \infty[)\,, \qquad \boldsymbol{y}|_{I_0} = \boldsymbol{\alpha}(x)\,, \tag{6), (7}$$

$$\boldsymbol{y}'(x) = A\boldsymbol{y}(\lambda x)\,, \quad \text{für } x \geqq x_0. \tag{8}$$

Im Punkt $x_0$ ist natürlich die rechtsseitige Ableitung gemeint. Existiert eine Lösung, so kann diese durch schrittweises Fortsetzen ($\lambda < 1$) der vorgegebenen Anfangsfunktion $\boldsymbol{\alpha}(x)$ gewonnen werden, so daß man eine eindeutige, für alle $x \geqq x_0$ definierte, mit wachsendem $x$ glatter werdende Lösung des AWP erhält.

In Teil I wird vorerst die Struktur aller ganzen Lösungen des AWP (3), (4) angegeben, falls die Matrix $A$ die Form eines JORDAN-Blocks besitzt. Anschließend wird ein System von Matrizen $F_n(x)$, $n \in Z$, bestimmt, deren Spaltenvektoren für $x \neq 0$ analytische Lösungen von (3) sind. Im Teil II wird gezeigt (vgl. Satz 6), daß sich jede Lösung des AWP (3), (5) mit Hilfe dieser Matrizen darstellen läßt. Die Konstruktion der $F_n(x)$ beruht auf der

Anpassung eines Verfahrens, das in [1] zur Lösung der skalaren Gleichung $F'(t) = e^{\alpha t+\beta}F(t-1)$ verwendet wurde; u. zw. wird durch einen Lösungsansatz in Form eines FOURIER-LAPLACE-Integrals (vgl. auch [4]) die Lösung der Funktional-Differentialgleichung auf die Lösung einer Differenzengleichung zurückgeführt. Diese Vorgangsweise wird durch die Substitution $\alpha t = \ln x$, mit der die in [1] untersuchte Gleichung auf die Form $y'(x) = ay(x/e)$ transformiert werden kann, nahegelegt. Außerdem hat sie den Vorteil, daß die in [1] angegebenen asymptotischen Beziehungen nach entsprechender Modifikation auch für die Abschätzung der $F_n(x)$ verwendet werden können (vgl. **3**).

Im Teil II wird vorerst mit Hilfe einer GREENschen Matrix $\Phi(x, t)$ eine Darstellung beliebiger Lösungen durch die Anfangsfunktion $\alpha(x)$ angegeben (vgl. (4.5)). Anschließend wird eine Reihendarstellung (Satz 5) für $\Phi(x, t)$ mit Hilfe der Matrizen $F_n(x)$, $G_n(t)$ bewiesen, woraus sich die bereits erwähnte Entwicklung beliebiger Lösungen nach den $F_n(x)$ ergibt. Die Zeilenvektoren der Matrizen $G_n(t)$ sind dabei Lösungen des zu (3) adjungierten Systems. Schließlich wird durch geeignete Vergleichsfunktionen das asymptotische Verhalten aller Lösungen charakterisiert (vgl. Satz 7 und 8). Da eine Anzahl von Beweisschritten ähnlich wie im skalaren Fall verläuft, werden einzelne Beweise nicht in allen Details durchgeführt.

## 1. Die ganzen Lösungen des AWP (3), (4)

Die Ermittlung der ganzen Lösungen des Systems (3) kann auf die gleiche Art wie bei Systemen gewöhnlicher Differentialgleichungen geschehen. Die Rolle der Exponentialfunktion $e^{tx}$ übernimmt die ganze Funktion (der Ordnung Null)

$$w(tx) = \sum_{n=0}^{\infty} \frac{\lambda^{\frac{n(n-1)}{2}}}{n!} (tx)^n = \sum_{n=0}^{\infty} p_n(x)\, t^n\,, \qquad p_0(x) = 1\,. \tag{1.1}$$

Wie man sofort nachprüft, genügt sie der Funktionalgleichung $w_x(tx) = tw(\lambda tx)$. Für die skalare Gleichung $y'(x) = ay(\lambda x)$ ist daher

$$w(ax) = \sum_{n=0}^{\infty} p_n(x)\, a^n\,, \qquad w(0) = 1\,, \tag{1.2}$$

eine ganze Lösung. Jede weitere ganze Lösung hat dann die Form $y(x) = y(0)\, w(ax)$. Die Reihe

$$W(Ax) = \sum_{n=0}^{\infty} p_n(x)\, A^n \tag{1.3}$$

konvergiert für beliebige Matrizen $A$ und stellt daher eine ganze Matrizenfunktion dar, die dem AWP

$$W'(Ax) = AW(\lambda Ax)\,, \qquad W(0) = E\,, \qquad E \text{ Einheitsmatrix}\,, \tag{1.4}$$

genügt. Wie üblich wird $W(Ax)$ als die, zum System (3) gehörende *Fundamentalmatrix* bezeichnet, da jede ganze Lösung des AWP (3), (4) offenbar die Form

$$\boldsymbol{y}(x) = W(Ax)\, \boldsymbol{y}_0 \tag{1.5}$$

hat. Sei nun $T$ eine reguläre Matrix mit deren Hilfe $A$ auf die JORDANsche Normalform (NF) transformiert werden kann. Durch die Transformation $\boldsymbol{y} = T\boldsymbol{u}$ erhält man für (3) das System

$$\boldsymbol{u}'(x) = J\boldsymbol{u}(\lambda x) = T^{-1}AT\boldsymbol{u}(\lambda x)\,. \tag{1.6}$$

Außerdem folgt aus $A^n = TJ^nT^{-1}$

$$W(Ax) = TW(Jx)\, T^{-1}\,, \tag{1.7}$$

d. h. es genügt, die Struktur der Matrix $W(Ax)$ für Systemmatrizen $A$ in Form eines JORDAN-Blocks zu untersuchen. Bekanntlich hat dann $A$ die Gestalt

$$A = \omega E + N \tag{1.8}$$

mit der nilpotenten $r \times r$-Matrix $N = \begin{pmatrix} 0 & 1 & & \\ & \ddots & \ddots & \\ & & \ddots & 1 \\ & & & 0 \end{pmatrix}$ und dem einzigen Eigenwert $\omega$. Für $A^n$ erhält man bei Berücksichtigung von $N^{r+k} = 0$, für $k \in \mathbb{N}_0$,

$$A^n = (\omega E + N)^n = \sum_{j=0}^{r-1} \binom{n}{j} \omega^{n-j} N^j\,,$$

woraus unmittelbar

$$W(Ax) = \begin{pmatrix} w(\omega x) & p_1(x)\, w(\lambda\omega x) & \dots & p_{r-1}(x)\, w(\lambda^{r-1}\omega x) \\ 0 & w(\omega x) & \dots & p_{r-2}(x)\, w(\lambda^{r-2}\omega x) \\ & & \dots & \\ 0 & 0 & \dots & w(\omega x) \end{pmatrix} \tag{1.9}$$

folgt. Es gilt daher

Satz 1: *Hat die Matrix $A$ die Form (1.8), dann ist jede ganze Lösung des Systems (3) eine Linearkombination von $r$ Vektoren $\boldsymbol{y}_k(x)$ der Form*

$$y_{kj}(x) = \begin{cases} p_{k-j}(x)\, w(\lambda^{k-j}\omega x)\,, & j \leqq k \\ 0, & j > k \end{cases}, \qquad k = 1, 2, \dots, r\,. \tag{1.10}$$

*Dabei sind die $p_{k-j}(x)$ die Polynome aus (1.1). Ist insbesondere $\omega = 0$, dann hat System (3) nur Polynomlösungen.*

## 2. Ein System spezieller analytischer Lösungen

Durch Satz 1 wurden alle ganzen Lösungen des Systems (3) charakterisiert. Schränkt man den Definitionsbereich der Lösungen geeignet ein, so besitzt (3) weitere analytische Lösungen, die vor allem im Hinblick auf das AWP (3), (5) von Interesse sind. Im weiteren Verlauf der Arbeit werden nur mehr Lösungen des Systems (3) für $x \neq 0$ bzw. $x \in \mathbb{R}_+$ betrachtet. Dabei kann man sich wieder auf $r \times r$-Matrizen $A$ in Form eines JORDAN-Blocks beschränken. Außerdem erweist sich die Einführung des zu (3) adjungierten Systems

$$\boldsymbol{g}'(x) = -\mu \boldsymbol{g}(\mu x)\, A\,, \qquad \mu = \lambda^{-1} > 1\,, \qquad x > 0\,, \tag{2.1}$$

als äußerst nützlich. Als Lösung von (2.1) wird jede Funktion $\boldsymbol{g}(x) = (g_1(x), \dots, g_r(x)) \in C(]0, \mu x_1])$, $x_1 \in \mathbb{R}_+$, die für $x \leq x_1$ dem System (2.1) genügt, bezeichnet.

Anmerkung: Existiert für eine vorgegebene Anfangsfunktion $\beta(x) \in C(I_1)$, $I_1 = [x_1, \mu x_1]$, eine Lösung von (2.1), dann kann diese wieder (eindeutig) durch schrittweises Fortsetzen nach links bestimmt werden.

Das folgende Lemma ermöglicht nun die Einführung eines inneren Produktes, das auf den Lösungsräumen von (3) und (2.1) definiert ist.

Lemma 1: *Seien $x_0, x_1 \in \mathbb{R}_+$, $x_0 < x_1$, fest gewählt und $\boldsymbol{y}(x)$ für $x \geqq x_0$, $\boldsymbol{g}(x)$ für $x \leqq x_1$ Lösungen von (3) bzw. (2.1), dann ist die für $x \in [x_0, x_1]$ definierte Funktion*

$$\langle \boldsymbol{g}(x), \boldsymbol{y}(x) \rangle = \boldsymbol{g}(x)\, \boldsymbol{y}(x) + \mu \int_{\lambda x}^{x} \boldsymbol{g}(\mu s)\, A \boldsymbol{y}(s)\, \mathrm{d}s\,, \qquad \boldsymbol{g}\boldsymbol{y} = \sum_{j=1}^{r} g_j y_j\,, \tag{2.2}$$

*unabhängig von $x$, d. h. konstant.*

Der Beweis folgt unmittelbar durch Differenzieren unter Benützung von (3) und (2.1).

Anmerkung: Die Bilinearform $\langle \boldsymbol{g}(x), \boldsymbol{y}(x) \rangle$ kann als inneres Produkt der Lösungen $\boldsymbol{y}(x)$, $\boldsymbol{g}(x)$ aufgefaßt werden. Anschließend werden $r \times r$-Matrizen $F(x)$, $G(x)$ untersucht, die den Matrizengleichungen (2.3) und (2.4) genügen. Funktionen der Form $\langle G(x), F(x) \rangle$, $\langle \boldsymbol{g}(x), F(x) \rangle$, $\langle G(x), \boldsymbol{y}(x) \rangle$ sind dann ebenfalls von $x$ unabhängig, d. h. konstante Matrizen bzw. Vektoren (vgl. z. B. (3.15) und (5.7)).

Nach diesen Vorbereitungen werden nun zwei Systeme von Matrizen $F_n(x)$, $G_n(x)$, $n \in \mathbb{Z}$, mit analytischen Elementen $f_{n;jk}(x)$, $g_{n;jk}(x)$ so bestimmt, daß sie den Matrizengleichungen

$$F_n'(x) = A F_n(\lambda x)\,, \qquad G_n'(x) = -\mu G_n(\mu x)\, A \tag{2.3), (2.4}$$

für $n \in \mathbb{Z}$ und $x \neq 0$, genügen. Betrachtet man zunächst die Matrix

$$F_0(x) = \int_{\mathbb{R}} \frac{x^t \lambda^{\frac{t(t-1)}{2}}}{\Gamma(t+1)}\, C(t)\, \mathrm{d}t\,, \tag{2.5}$$

so sieht man durch formales Differenzieren unter dem Integral, daß $F_0(x)$ eine Lösung von (2.3) ist, falls die Matrix $C(t)$ der Differenzengleichung

$$C(t+1) = A C(t) \tag{2.6}$$

genügt. Setzt man

$$C(t) = \omega^t B(t) \tag{2.7}$$

und fordert, daß $B(t)$ zusätzlich eine obere Dreiecksmatrix mit der Normierungsbedingung $\operatorname{Det} B(t) = 1$ ist, so folgt aus (2.6) unmittelbar

$$\omega[B(t+1) - B(t)] = N B(t)\,. \tag{2.8}$$

System (2.8) läßt sich leicht lösen, und man erhält

$$B(t) = \begin{pmatrix} N_0(t) & \omega^{-1} N_1(t) & \dots & \omega^{1-r} N_{r-1}(t) \\ 0 & N_0(t) & \dots & \omega^{2-r} N_{r-2}(t) \\ & \dots & & \\ 0 & 0 & \dots & N_0(t) \end{pmatrix}. \tag{2.9}$$

Dabei wird mit $N_j(t) = \binom{t}{j} = \dfrac{t(t-1)\dots(t-j+1)}{j!}$, $j \leqq r-1$, das $j$-te NEWTONsche Interpolationspolynom bezeichnet.

Anmerkung: Die Folge $\{N_j(t)\}$, $j \in \mathbb{N}_0$, genügt bekanntlich der Differenzengleichung

$$N_j(t+1) - N_j(t) = N_{j-1}(t)\,, \qquad N_0 = 1\,, \tag{2.10}$$

mit deren Hilfe man direkt abliest, daß $B(t)$ eine Lösung von (2.8) ist.

Mit (2.5) erhält man daher endgültig

$$F_0(x) = \int\limits_{\mathbb{R}} \frac{(\omega x)^t \lambda^{\frac{t(t-1)}{2}}}{\Gamma(t+1)} B(t)\,\mathrm{d}t\,. \tag{2.11}$$

In $(\omega x)^t = e^{t \ln \omega x}$ ist dabei der Hauptwert einzusetzen. Die Matrix $F_0(x)$ ist für $x \neq 0$ analytisch und eine Lösung von (2.3). Natürlich sind die Matrizen

$$F_n(x) = \int\limits_{\mathbb{R}} \frac{(\omega x)^t \lambda^{\frac{t(t-1)}{2}} e^{2\pi i n t}}{\Gamma(t+1)} B(t)\,\mathrm{d}t\,, \qquad n \in \mathbb{Z}\,, \tag{2.12}$$

ebenfalls Lösungen von (2.3).

Anmerkung: Man erhält $F_n(x)$ aus $F_0(x)$, wenn man in (2.11) den $n$-ten Zweig des Logarithmus wählt. Setzt man $F_0(x) = \tilde{F}_0(\ln x)$, so kann man

$$F_n(x) = \tilde{F}_0\,(\ln x + 2\pi i n) \tag{2.13}$$

schreiben. Im weiteren Verlauf der Arbeit wird mit $\ln x$ immer der Hauptwert bezeichnet. Außerdem sind Ausdrücke der Form $a^b$ immer als $e^{b \ln a}$ zu interpretieren.

Eine analoge Konstruktion führt auch beim adjungierten System (2.4) zum Ziel. Mit dem Lösungsansatz

$$G_n(x) = (2\pi i)^{-1} \int\limits_{1-i\infty}^{1+i\infty} x^{-t} \mu^{\frac{t(t-1)}{2}} e^{-2\pi i n t}\, \Gamma(t)\, C^*(t)\,\mathrm{d}t\,, \qquad \mu = \lambda^{-1} > 1,\ n \in \mathbb{Z}\,, \tag{2.14}$$

erhält man zunächst formal die Bedingung

$$C^*(t-1) = C^*(t)\,A\,. \tag{2.15}$$

Wegen $\mu > 1$ ist der Integrationsweg so zu wählen, daß (2.14) analytische Lösungen liefert. Der Faktor $(2\pi i)^{-1}$ dient nur Normierungszwecken.

Setzt man analog zum vorhergehenden Fall

$$C^*(t) = \omega^{-t} B^*(t)\,, \tag{2.16}$$

wobei $B^*(t)$ wieder eine obere Dreiecksmatrix mit der Normierung $\operatorname{Det} B^*(t) = 1$ sein soll, so erhält man unmittelbar

$$B^*(t) = \begin{pmatrix} 1 & \omega^{-1}\check{N}_1(t) & \dots & \omega^{1-r}\check{N}_{r-1}(t) \\ 0 & 1 & \dots & \omega^{2-r}\check{N}_{r-2}(t) \\ & \dots & & \\ 0 & 0 & \dots & 1 \end{pmatrix}. \tag{2.17}$$

Die Polynome $\check{N}_j(t) = N_j(-t) = \frac{(-1)^j}{j!} \prod\limits_{k=0}^{j-1} (t+k)$ genügen der Funktionalgleichung

$$\check{N}_j(t-1) - \check{N}_j(t) = \check{N}_{j-1}(t)\,, \qquad \check{N}_0 = 1\,. \tag{2.18}$$

Mit (2.14) ergibt sich daher das folgende System analytischer Lösungen der adjungierten Gleichung (2.4)

$$G_n(x) = (2\pi i)^{-1} \int\limits_{1-i\infty}^{1+i\infty} (\omega x)^{-t} \mu^{\frac{t(t-1)}{2}} e^{-2\pi i n t}\, \Gamma(t)\, B^*(t)\,\mathrm{d}t\,, \qquad n \in \mathbb{Z}\,, \quad x \neq 0\,. \tag{2.19}$$

Die Matrizen $F_n(x)$, $G_n(x)$ lassen sich noch so umformen, daß die in [1] für den skalaren Fall angegebenen asymptotischen Abschätzungen verwendet werden können.

Satz 2: *Seien*

$$f_n(x) = f_{n;jj}(x) = \int\limits_{\mathbb{R}} \frac{(\omega x)^t \lambda^{\frac{t(t-1)}{2}} e^{2\pi i n t}}{\Gamma(t+1)}\,\mathrm{d}t \tag{2.20}$$

bzw.

$$g_n(x) = g_{n;jj}(x) = (2\pi i)^{-1} \int\limits_{1-i\infty}^{1+i\infty} (\omega x)^{-t} \mu^{\frac{t(t-1)}{2}} e^{-2\pi i n t}\, \Gamma(t)\,\mathrm{d}t\,, \qquad n \in \mathbb{Z}\,, \qquad j = 1, 2, \dots, r\,, \tag{2.21}$$

*die Elemente der Hauptdiagonale der Matrizen $F_n(x)$ bzw. $G_n(x)$ und $p_j(x)$, $q_j(x)$ die Polynome*

$$p_j(x) = \lambda^{\frac{j(j-1)}{2}} \frac{x^j}{j!}, \qquad q_j(x) = \mu^{\frac{j(j-1)}{2}} \frac{(-x)^j}{j!}, \tag{2.22}$$

*dann gilt*

$$f_{n;jk}(x) = p_{k-j}(x)\, f_n(\lambda^{k-j}x) \quad bzw. \quad g_{n;jk}(x) = q_{k-j}(\mu x)\, g_n(\mu^{k-j}x) \quad \text{für } j \leqq k \leqq r \text{ und } n \in \mathbb{Z}\,. \tag{2.23}$$

Anmerkung: Man prüft sofort nach, daß die Funktionen (2.20), (2.21) den skalaren Gleichungen

$$f_n'(x) = \omega f_n(\lambda x)\,, \qquad g_n'(x) = -\omega\mu g_n(\mu x) \tag{2.24}$$

genügen. Setzt man noch (vgl. (2.13)) $f_0(x) = \tilde{f}_0(\ln x)$, $g_0(x) = \tilde{g}_0(\ln x)$, so folgt

$$f_n(x) = \tilde{f}_0(\ln x + 2\pi i n)\,, \qquad g_n(x) = \tilde{g}_0(\ln x + 2\pi i n)\,. \tag{2.25}$$

Beweis: Offenbar genügt es, (2.23) für $n = 0$ zu zeigen.

$$f_{0;jk}(x) = \int\limits_{\mathbb{R}} \frac{(\omega x)^t \lambda^{\frac{t(t-1)}{2}} N_{k-j}(t)}{\omega^{k-j}\Gamma(t+1)}\, dt = \frac{\lambda^{\frac{(k-j)(k-j-1)}{2}} x^{k-j}}{(k-j)!} \int\limits_{\mathbb{R}} \frac{(\lambda^{k-j}\omega x)^t \lambda^{\frac{t(t-1)}{2}}}{\Gamma(t+1)}\, dt = p_{k-j}(x)\, f_0(\lambda^{k-j}x)\,.$$

Die Darstellung der $g_{0;jk}(x)$ ergibt sich ebenfalls unmittelbar, wenn man berücksichtigt, daß der Integrationsweg in (2.19) beliebig nach rechts (parallel) verschoben werden kann, ohne daß sich der Wert des Integrals ändert.

Anmerkung: Die Elemente $w_{jk}(x)$ der Fundamentalmatrix $W(Ax)$ haben gemäß (1.9) die Form $w_{jk}(x) = p_{k-j}(x)\, w(\lambda^{k-j}\omega x)$, $j \leqq k \leqq r$, mit $w(\lambda^{k-j}\omega x) = \sum\limits_{n=0}^{\infty} \frac{\lambda^{\frac{n(n-1)}{2}}}{n!} (\lambda^{k-j}\omega x)^n$. Nach Satz 2 gilt

$$f_{n;jk}(x) = p_{k-j}(x) \int\limits_{\mathbb{R}} \frac{(\lambda^{k-j}\omega x)^t \lambda^{\frac{t(t-1)}{2}} e^{2\pi i n t}}{\Gamma(t+1)}\, dt\,,$$

woraus mit Hilfe der Poissonschen Summenformel

$$W(Ax) = \sum_{n=-\infty}^{\infty} F_n(x)\,, \qquad x \neq 0\,, \tag{2.26}$$

folgt, d. h. die Fundamentalmatrix $W(Ax)$ der ganzen Lösungen läßt sich als Reihe der $F_n(x)$ darstellen. In 5 wird gezeigt, daß dies für jede Lösung des AWP (3), (5) der Fall ist.

## 3. Asymptotisches Verhalten und Biorthogonalität der Matrizen $F_n(x)$, $G_n(x)$

Da die in den Matrizen $F_n(x)$, $G_n(x)$ auftretenden Funktionen $f_n(\lambda^{k-j}x)$, $g_n(\mu^{k-j}x)$ den skalaren Funktionalgleichungen

$$\frac{d}{dx} f_n(\lambda^{k-j}x) = \omega\lambda^{k-j} f_n(\lambda^{k-j+1}x) \quad \text{bzw.} \quad \frac{d}{dx} g_n(\mu^{k-j}x) = -\omega\mu^{k-j+1} g_n(\mu^{k-j+1}x) \tag{3.1), (3.2}$$

genügen, können die in [1] durch Anwendung von Sattelpunktmethoden auf die Integrale (2.20), (2.21) erhaltenen asymptotischen Beziehungen (vgl. insbesondere [1], (1.2), (5.16), (5.17), (5.19)) auch zur Abschätzung des Wachstums der Matrizen $F_n(x)$, $G_n(x)$ bzw. der durch sie bestimmten Lösungen verwendet werden. Die folgenden asymptotischen Formeln gelten für $x \in \mathbb{R}_+$ und $x \to \infty$, oder gleichmäßig in jeder festen Halbebene $\operatorname{Re} x \geqq x_0 > 0$, was durch $|x| \to \infty$ gekennzeichnet werden soll. Ist $h(\ln x)$ die Funktion

$$h(\ln x) = \frac{1}{2c}(\ln x - \ln\ln x)^2 + \left(\frac{1}{2} + \frac{1}{c} + \frac{\ln \omega c}{c}\right)\ln x - \left(1 + \frac{\ln \omega c}{c}\right)\ln\ln x + \left(1 + \frac{\ln \omega c}{c}\right)\frac{\ln\ln x}{\ln x} - \frac{\ln^2 \ln x}{2c \ln x} - \left(1 + \frac{\ln \omega^2 c}{2c}\right)\frac{\ln c}{\ln x} + d \tag{3.3}$$

mit

$$c = |\ln \lambda|\,, \qquad d = \frac{\ln c}{2c}(c + \ln \omega^2 c) + \frac{1}{2}\left(\frac{c}{4} + \ln \omega + \frac{\ln^2 \omega}{c}\right), \tag{3.4}$$

dann gelten folgende asymptotische Formeln, die durch die Transformation $ct = \ln x$ aus dem in [1] angegebenen asymptotischen Verhalten der Lösungen folgen.

$$f_0(x) = e^{h(\ln x)}\left[1 + O\left(\frac{1}{\ln x}\right)\right], \qquad |x| \to \infty; \tag{3.5}$$

$$g_0(x) = \frac{e^{-h(\ln x)}}{\ln x}\left[1 + \frac{\ln\ln x}{\ln x} + O\left(\frac{1}{\ln x}\right)\right], \qquad |x| \to \infty\,; \tag{3.6}$$

$$f_0(\lambda^k x) = f_0(x) \exp\left\{\frac{ck^2}{2}\left(1 - \frac{2}{\ln x}\right) - k\left[\ln x - \ln\ln x + \frac{\ln\ln x}{\ln x} + \gamma - (c + \ln \omega c)\frac{1}{\ln x}\right] + o\left(\frac{1}{\ln x}\right)\right\}$$

$$= f_0(x) \exp\left\{\frac{ck^2}{2} - k\left[\ln x - \ln\ln x + \frac{\ln\ln x}{\ln x} + \gamma\right] + O\left(\frac{1}{\ln x}\right)\right\}, \gamma = \frac{c}{2} + \ln \omega c, \quad k \in \mathbb{N}_0, \quad |x| \to \infty; \tag{3.7}$$

$$f_n(\lambda^k x) = f_0(\lambda^k x) \exp\left\{-\frac{2\pi^2 n^2}{c} + \frac{2\pi i n}{c}\left[\ln x - \ln\ln x + \frac{\ln\ln x}{\ln x} + \gamma\right] + O\left(\frac{1}{\ln x}\right)\right\}, \tag{3.8}$$

für festes $n \in \mathbb{Z}$ und $x \to \infty$.

Anmerkung: Für festes $\Delta$ bzw. $\Delta = O[(\ln x)^\beta]$, $\beta < \frac{1}{2}$, (im folgenden $o$-Term treten Summanden der Form $\frac{\Delta^2}{\ln^2 x}$ auf) folgt mit (3.3)

$$h(\ln x + \Delta) = h(\ln x) + \frac{\Delta}{c}\left[\ln x - \ln\ln x + \frac{\ln\ln x}{\ln x} + \gamma - (c + \ln c)\frac{1}{\ln x}\right] + \frac{\Delta^2}{2c}\left(1 - \frac{2}{\ln x}\right) + o\frac{1}{\ln x}\Big),$$

$$|x| \to \infty, \tag{3.9}$$

woraus man mit $\Delta = 2\pi i n$ bzw. $\Delta = -ck$ direkt (3.7) und (3.8) abliest.

Für die $g_m(x)$ ergibt sich mit (3.6) und (3.9)

$$g_0(\mu^j x) = g_0(x) \exp\left\{-\frac{cj^2}{2} - j\left[\ln x - \ln\ln x + \frac{\ln\ln x}{\ln x} + \gamma\right] + O\left(\frac{1}{\ln x}\right)\right\}, \quad j \in \mathbb{N}_0, \quad |x| \to \infty; \tag{3.10}$$

$$g_m(\mu^j x)\ g_0(\mu^j x) \exp\left\{\frac{2\pi^2 m^2}{c} - \frac{2\pi i m}{c}\left[\ln x - \ln\ln x + \frac{\ln\ln x}{\ln x} + \gamma\right] + O\left(\frac{1}{\ln x}\right)\right\}, \ m \in \mathbb{Z},\ x \to \infty. \tag{3.11}$$

Ist andererseits $I \subset \mathbb{R}_+$ ein beliebiges, kompaktes Intervall, dann gelten für $|n| \to \infty$ und $x \in I$ außerdem folgende Beziehungen

$$f_n(\lambda^k x) = f_n(1) \exp\left\{\frac{\ln \lambda^k x}{c}\left[2\pi i n - \ln 2\pi |n| - \frac{i\pi}{2}\operatorname{sgn} n + \gamma\right] + \frac{\ln^2 \lambda^k x}{2c}\left(1 - \frac{1}{\pi i n}\right) + O\left(\frac{\ln |n|}{|n|}\right)\right\}, \tag{3.12}$$

$$k \in \mathbb{N}_0, \qquad f_n(1) = \tilde{f}_0(\ln 1 + 2\pi i n), \qquad \text{(vgl. (2.27))};$$

$$g_n(\mu^j x) = g_n(1) \exp\left\{-\frac{\ln \mu^j x}{c}\left[2\pi i n - \ln 2\pi |n| - \frac{i\pi}{2}\operatorname{sgn} n + \gamma\right] - \frac{\ln^2 \mu^j x}{2c}\left(1 - \frac{1}{\pi i n}\right) - O\left(\frac{\ln |n|}{|n|}\right)\right\},$$

$$j \in \mathbb{N}_0; \tag{3.13}$$

$$f_n(1)\, g_n(1) = \frac{1}{2\pi i n}\left[1 + O\left(\frac{\ln |n|}{|n|}\right)\right]. \tag{3.14}$$

Mit Hilfe dieser asymptotischen Beziehungen läßt sich nun der folgende Satz beweisen.

Satz 3: *Die Matrizen $F_n(x)$, $G_m(x)$, $m, n \in \mathbb{Z}$, genügen für $x \in \mathbb{R}_+$ der Beziehung*

$$\langle G_m(x), F_n(x)\rangle = G_m(x)\, F_n(x) + \mu \int_{\lambda x}^{x} G_m(\mu t)\, A F_n(t)\, dt = E_{mn} = \begin{cases} E, & m = n \\ 0, & m \neq n \end{cases}, \quad 0\text{: } \textit{Nullmatrix}. \tag{3.15}$$

Anmerkung: Aus (3.15) folgt insbesondere die Biorthogonalität der $l$-ten Zeilen- und Spaltenvektoren $l = 1, 2, \ldots, r$, der Matrizen $G_m(x)$ und $F_n(x)$ bezüglich des inneren Produktes (2.2). Der Einfachheit halber wird daher auch vom „Biorthogonalsystem" $\langle G_m(x), F_n(x)\rangle$ gesprochen.

Beweis: Offenbar sind die Matrizen $E_{mn}$ konstante obere Dreiecksmatrizen, deren Elemente $e_{mn;jk}$, $k \geqq j$, mit den Bezeichnungen aus Satz 2 für $A = \omega E + N$ die Form

$$e_{mn;jk} = \sum_{l=j}^{k} g_{m;jl}(x)\, f_{n;lk}(x) + \mu\omega \sum_{l=j}^{k} \int_{\lambda x}^{x} g_{m;jl}(\mu t)\, f_{n;lk}(t)\, dt + \mu \sum_{l=1}^{k-1} \int_{\lambda x}^{x} g_{m;jl}(\mu t)\, f_{n;(l+1)k}(t)\, dt \tag{3.16}$$

besitzen. Für $k = j$ erhält man daher

$$e_{mn;jj} = g_m(x)\, f_n(x) + \mu\omega \int_{\lambda x}^{x} g_m(\mu t)\, f_n(t)\, dt\,. \tag{3.17}$$

Im weiteren Verlauf werden nun 2 Fälle unterschieden. Der Fall $j < k$ läßt sich leicht direkt erledigen, während für den Fall $j = k$ die asymptotischen Abschätzungen (3.5) bis (3.11) herangezogen werden.

1. $e_{mn;jk} = 0$ für $j < k \leqq r$ und $m, n \in \mathbb{Z}$: Wegen der speziellen Form der Matrizen $G_m(x)$, $F_n(x)$ kann man sich auf die Elemente $e_{mn;1k}$ der ersten Zeile beschränken. Außerdem ist der Beweis von $m$ und $n$ unabhängig. Mit Hilfe von (2.23) und den Funktionalgleichungen (2.24) lassen sich die zwei letzten Terme in (3.16) auf die Form

$$\frac{\mu}{\omega^{k-2}} \int_{\lambda x}^{x} \frac{t^{k-1}}{(k-1)!} [g_m(\mu t) f_n(t)]^{(k-1)} \, dt \quad \text{bzw.} \quad \frac{\mu}{\omega^{k-2}} \int_{\lambda x}^{x} \frac{t^{k-2}}{(k-2)!} [g_m(\mu t) f_n(t)]^{(k-2)} \, dt$$

bringen, woraus durch partielle Integration sofort die Behauptung folgt.

2. $e_{mn;jj} = \begin{cases} 1, & m = n \\ 0, & m \neq n \end{cases}$, $j = 1, 2, \dots, r$. Für $x \to \infty$ ergibt sich mit (3.5) und (3.6)

$$g_0(x) f_0(x) = \frac{1}{\ln x} [1 + o(1)] = o(1) , \tag{3.18}$$

woraus mit (3.8) und (3.11)

$$g_m(x) f_n(x) = o(1) , \qquad m, n \in \mathbb{Z} , \tag{3.19}$$

folgt. Außerdem erhält man bei Berücksichtigung von (3.8), (3.10), (3.11) und (3.18)

$$\mu\omega g_m(\mu x) f_n(x) = \frac{1}{cx} \exp\left\{\frac{2\pi^2}{c} (m^2 - n^2) + \frac{2\pi i}{c} (n - m) [\ln x - \ln \ln x + \gamma] + o(1)\right\}. \tag{3.20}$$

Für $n = m$ folgt mit (3.20) und $c = |ln\, \lambda|$

$$\mu\omega \int_{\lambda x}^{x} g_n(\mu t) f_n(t) \, dt = \frac{e^{o(1)}}{c} \int_{\lambda x}^{x} \frac{dt}{t} = e^{o(1)} .$$

Nun ist (3.17) das innere Produkt für den skalaren Fall und daher eine Konstante. Mit (3.19) ergibt sich daher $e_{nn;jj} = 1$. Für $n \neq m$ folgt mit (3.17), (3.19) und (3.20) $e_{mn;jj} = 0$, $j = 1, 2, \dots, r$, falls noch

$$\int_{\lambda x}^{x} \frac{\exp\{2\pi i(n - m) c^{-1}(\ln t - \ln \ln t)\}}{t} \, dt = o(1) , \qquad x \to \infty ,$$

berücksichtigt wird.

## TEIL II: DARSTELLUNG UND ASYMPTOTISCHES VERHALTEN ALLER LÖSUNGEN

### 4. Die Greensche Matrix $\Phi(x, t)$

Das Ziel des zweiten Teiles ist die Darstellung beliebiger Lösungen des AWP (3), (5) mit Hilfe des Biorthogonalsystems $\langle G_m(x), F_n(x)\rangle$ und die anschließende Abschätzung des asymptotischen Verhaltens dieser Lösungen. Dazu wird in 4 vorerst eine Darstellung beliebiger Lösungen mit Hilfe der Anfangsfunktion $\boldsymbol{\alpha}(x)$ und einer Greenschen Matrix $\Phi(x, t)$ angegeben.

Sei $x \in \mathbb{R}_+$ fest gewählt und die $r \times r$-Matrix $\Phi^*(t, x)$ als Funktion von $t$ für $t \leqq x$ eine Lösung der adjungierten Gleichung (2.4) mit

$$\Phi^*(t, x) = \begin{cases} 0, & t > x \\ E, & t = x. \end{cases} \tag{4.1}$$

Anmerkung: Die Matrix $\Phi^*(t, x)$ läßt sich leicht durch schrittweises Fortsetzen (nach links) mit Hilfe von (2.4) bestimmen, wobei man für $t \in [\lambda x, x]$ von $\frac{d}{dt} \Phi^*(t, x) = -\mu\Phi^*(\mu t, x) A = 0$ ausgeht.

Mit (3) und (4.1) folgt nach einfacher Zwischenrechnung die Darstellung

$$\boldsymbol{y}(x) = \Phi^*(\xi, x) \boldsymbol{y}(\xi) + \mu \int_{\lambda\xi}^{\xi} \Phi^*(\mu t, x) A\boldsymbol{y}(t) \, dt , \qquad 0 < \xi \leqq x . \tag{4.2}$$

Ist nun die $r \times r$-Matrix $\Phi(x, \xi)$ für $x \geqq \xi$ eine Lösung von (2.3), die der Bedingung

$$\Phi(x, \xi) = \begin{cases} 0, & x < \xi \\ E, & x = \xi \end{cases} \tag{4.3}$$

genügt, so ergibt sich aus (4.2) unmittelbar

$$\Phi(x, \xi) = \Phi^*(\xi, x) . \tag{4.4}$$

Die gewünschte Darstellung der Lösungen des AWP (3), (5) hat daher die Form

$$\boldsymbol{y}(x) = \Phi(x, x_0) \boldsymbol{\alpha}(x_0) + \mu \int_{\lambda x_0}^{x_0} \Phi(x, \mu t) A\boldsymbol{\alpha}(t) \, dt . \tag{4.5}$$

Anmerkung: Analog zeigt man für die Lösungen $\boldsymbol{g}(x)$, $x \leqq x_1$, der adjungierten Gleichung (2.1) die Darstellung

$$\boldsymbol{g}(x) = \boldsymbol{\beta}(x_1)\, \Phi(x_1, x) + \int_{x_1}^{\mu x_1} \boldsymbol{\beta}(t)\, A\Phi(\lambda t, x)\, dt \tag{4.6}$$

mit der vorgegebenen Anfangsfunktion $\boldsymbol{\beta}(x) \in C([x_1, \mu x_1])$, $x_1 \in R_+$.

Bezeichnet man mit $1_+(x)$ die Heavisidesche Sprungfunktion und mit $\delta(x - t)$ die in $t$ konzentrierte Dirac-Distribution, dann gilt

Satz 4: *Für $x, t \in R_+$ hat die Greensche Matrix $\Phi(x, t)$ die Form*

$$\Phi(x, t) = \sum_{n=0}^{\infty} p_n(x - \mu^n t)\, 1_+(x - \mu^n t)\, A^n \tag{4.7}$$

*mit* $p_n(x) = \lambda^{\frac{n(n-1)}{2}} \dfrac{x^n}{n!}$.

Anmerkung: Die Reihe in (4.7) ist wegen der Faktoren $1_+(x - \mu^n t)$, $\mu > 1$, für jedes $x \in R_+$ eine endliche Summe, die für $x < t$ identisch verschwindet und für $x = t$ in die Einheitsmatrix $E$ übergeht.

Der Beweis von Satz 4 ergibt sich direkt durch Differenzieren von (4.7) unter Benützung der für die Polynomfolge $\{p_n(x)\}$ gültigen Funktionalgleichung $p_n'(x) = p_{n-1}(\lambda x)$, sowie der Beziehung $f(x)\,\delta(x - t) = f(t)\,\delta(x - t)$.

Wegen $1_+(x - \mu^n t) = \begin{cases} 0, & x < \mu^n t, \\ 1, & x \geqq \mu^n t, \end{cases}$ kann man (4.7) auch ohne die Heavisidesche Funktion schreiben, u. zw. erhält man

$$\Phi(x, t) = \sum_{0 \leq n \leq \frac{1}{c} \ln \frac{x}{t}} p_n(x - \mu^n t)\, A^n\,, \qquad c = |\ln \lambda|\,. \tag{4.8}$$

Für $x < t$ enthält die Summe keinen Summanden und ist daher als die Nullmatrix 0 zu interpretieren. Für $x = t$ folgt $\Phi(t, t) = E$, $(p_0 = 1)$. Berücksichtigt man noch die spezielle Form der Matrizen $A^n$, so ergibt sich für die Elemente $\varphi_{jk}(x, t)$ von $\Phi(x, t)$

$$\varphi_{jk}(x, t) = \sum_{0 \leq n \leq \frac{1}{c} \ln \frac{\lambda^{k-j} x}{t}} \binom{n + k - j}{k - j} \omega^n p_{n+k-j}(x - \mu^{n+k-j} t) \qquad j \leqq k\,, \qquad k = 1, 2, \ldots, r\,. \tag{4.9}$$

Für $j > k$ ist natürlich $\varphi_{jk}(x, t) \equiv 0$. Außerdem gilt

$$\varphi_{jk}(x, t) = \begin{cases} 0 & \text{für} \quad \lambda^{k-j} x \leqq t \quad \text{und} \quad j < k \\ 1 & \text{für} \quad x = t \quad \text{und} \quad j = k\,. \\ 0 & \text{für} \quad x < t \quad \text{und} \quad j = k \end{cases} \tag{4.10}$$

Anmerkung: Wegen $p_n(0) = 0$, $n \geqq 1$, sind die $\varphi_{jk}(x, t)$ für $j < k$ stetige Funktionen.

## 5. Die Reihendarstellung beliebiger Lösungen

Zunächst wird diese Reihendarstellung für die Greensche Matrix $\Phi(x, t)$ gezeigt. Mit (4.5) wird dann anschließend die gewünschte Darstellung für beliebige (stetige) Lösungen bewiesen.

Satz 5: *Sind die $\varphi_{jk}(x, t)$ bzw. $\varphi_{n;jk}(x, t)$, $j, k = 1, 2, \ldots, r$, $n \in Z$, die Elemente der Matrizen $\Phi(x, t)$ bzw. $\Phi_n(x, t) = F_n(x)\, G_n(t)$, dann gilt:*

1. *Die Reihen*

$$\sum_{n=-\infty}^{\infty} \varphi_{n;jk}(x, t) = \sum_{n=-\infty}^{\infty} \sum_{l=j}^{k} p_{l-j}(x)\, q_{k-l}(\mu t)\, f_n(\lambda^{l-j} x)\, g_n(\mu^{k-l} t)$$

*konvergieren für $x > \lambda t$, wobei die Konvergenz für $x > t$ absolut ist.*

2. *Die $\varphi_{jk}(x, t)$ besitzen für $x > \mu^{k-j-1} t$ und $j < k$ bzw. für $x > t$, $\lambda t < x < t$ und $j = k$ die Darstellung*

$$\varphi_{jk}(x, t) = \sum_{n=-\infty}^{\infty} \varphi_{n;jk}(x, t)\,. \tag{5.1}$$

*Für $x = t$ und $j = k$ konvergiert die Reihe gegen das arithmetische Mittel $\frac{1}{2}[\varphi_{jj}(t, t + 0) + \varphi_{jj}(t, t - 0)] = \frac{1}{2}$.*

Anmerkung: Der Fall $j > k$ ist wegen $\varphi_{jk} \equiv \varphi_{n;jk} \equiv 0$ trivial. Außerdem bedeutet die Tatsache, daß die Gültigkeit von (5.1) nur für $x > \mu^{k-j-1} t$ bewiesen wird, keine Einschränkung, da wegen $\varphi_{jk}(x, t) = 0$, für $x \leqq \mu^{k-j} t$, $j < k$, (vgl. (4.10)) und Darstellung (4.5) die $\varphi_{jk}(x, t)$ erst für $x > \mu^{k-j} x_0$ einen Beitrag zur Lösung liefern. Setzt man daher für $j < k$ die Reihen in (5.1) durch Multiplikation mit den Heaviside-Funktionen $1_+(x - \mu^{k-j} t)$ auf $\lambda t < x \leqq \mu^{k-j-1} t$ stetig durch Null fort, so kann (5.1) auch in der Form

$$\Phi(x, t) = \sum_{n=-\infty}^{\infty} F_n(x)\, G_n(t)\,, \qquad x > \lambda t\,, \qquad x \neq t\,, \qquad t \in R_+\,, \tag{5.2}$$

geschrieben werden, die im Zusammenhang mit (4.5) (vgl. auch Satz 6) eine bequemere Schreibweise erlaubt.

Beweis:

1. Konvergenz: Für fest gewählte $x, t \in \mathbb{R}_+$, $u, v \in \mathbb{N}_0$ ergibt sich mit (3.12) bis (3.14) für $n \to \infty$

$$f_n(\lambda^u x)\, g_n(\mu^v t) = -\, i \exp\left\{\frac{1}{c} \ln \lambda^{u+v} \frac{x}{t}\left[2\pi i n - \frac{i\pi}{2} \operatorname{sgn} n + \gamma\right] + \right.$$

$$\left. + \frac{1}{2c} (\ln^2 \lambda^u x - \ln^2 \mu^v t)\left(1 - \frac{1}{i\pi n}\right)\right\} (2\pi |n|)^{\frac{1}{c} \ln \mu^{u+v} \frac{t}{x} - 1}\left[1 + O\, \frac{\ln |n|}{|n|}\right], \tag{5.3}$$

woraus für $k = j$

$$\varphi_{n;jj}(x, t) = -i \exp\left\{\frac{1}{c} \ln \frac{x}{t}\left(2\pi i n - \frac{i\pi}{2} \operatorname{sgn} n + \gamma\right) + \frac{1}{2c} (\ln^2 x - \ln^2 t)\right\} (2\pi |n|)^{\frac{1}{c} \ln \frac{t}{x} - 1}\left[1 + O\left(\frac{\ln |n|}{|n|}\right)\right] \tag{5.4}$$

folgt. Für $j < k$, $u = l - j$, $v = k - l$, erhält man aus (5.3) unter Berücksichtigung der speziellen Form der Polynome $p_{l-j}(x)$, $q_{k-l}(\mu t)$ mit $e^c = \mu$

$$\varphi_{n;jk}(x, t) = -\, i\mu^{\frac{(k-j)^2}{2}}\, p_{k-j}\left(x^{\frac{1}{i\pi n}} - [\mu^{k-j} t]^{\frac{1}{i\pi n}}\right) \times$$

$$\times \exp\left\{\frac{1}{c}\left(2\pi i n - \frac{i\pi}{2} \operatorname{sgn} n + \gamma\right) \ln \lambda^{k-j} \frac{x}{t} + \frac{1}{2c} (\ln^2 x - \ln^2 t)\right\} (2\pi |n|)^{\frac{1}{c} \ln \mu^{k-j} \frac{t}{x} - 1}\left[1 + O\left(\frac{\ln |n|}{|n|}\right)\right].$$

Da außerdem $p_{k-j}\left(x^{\frac{1}{i\pi n}} - [\mu^{k-j} t]^{\frac{1}{i\pi n}}\right) = O([2\pi |n|]^{j-k})$, $n \to \infty$, gilt, folgt endgültig

$$\varphi_{n;jk}(x, t) = b(n) \exp\left\{\frac{1}{c}\left(2\pi i n - \frac{i\pi}{2} \operatorname{sgn} n\right) \ln \lambda^{k-j} \frac{x}{t}\right\} (2\pi |n|)^{\frac{1}{c} \ln \frac{t}{x} - 1} \tag{5.5}$$

mit der (für festes $x$ und $t$) beschränkten Funktion

$$b(n) = -\, i\mu^{\frac{(k-j)^2}{2}}\, p_{k-j}\left(x^{\frac{1}{i\pi n}} - [\mu^{k-j} t]^{\frac{1}{i\pi n}}\right) (2\pi |n|)^{k-j} \times$$

$$\times \exp\left\{\frac{\gamma}{c} \ln \lambda^{k-j} \frac{x}{t} + \frac{1}{2c} (\ln^2 x - \ln^2 t)\right\}\left[1 + O\left(\frac{\ln |n|}{|n|}\right)\right].$$

Aus (5.4) und (5.5) folgt nun unmittelbar das im Satz angegebene Konvergenzverhalten.

Die Darstellung (5.1) kann nach entsprechender Modifikation ähnlich wie im skalaren Fall bewiesen werden. Dazu werden die $\varphi_{n;jk}(x, t)$ vorerst durch ein einziges Integral ausgedrückt, auf das die Poissonsche Summenformel angewendet werden kann (man vgl. dazu den in [1] für den skalaren Fall angegebenen Beweis).

Mit Satz 5 und der Integraldarstellung (4.5) erhält man nun die eingangs erwähnte Reihendarstellung beliebiger Lösungen des AWP (3), (5).

Satz 6: *Ist $\boldsymbol{y}(x)$ für $x \geqq x_0$, $x_0 \in \mathbb{R}_+$, eine Lösung des* AWP (3), (5), *dann gilt*

$$\boldsymbol{y}(x) = \sum_{n=-\infty}^{\infty} F_n(x) \langle G_n, \boldsymbol{y}\rangle \quad \text{für} \quad x > x_0 \tag{5.6}$$

*mit*

$$\boldsymbol{c}_n = \langle G_n, \boldsymbol{y}\rangle = G_n(x_0)\, \boldsymbol{y}(x_0) + \mu \int_{\lambda x_0}^{x_0} G_n(\mu t)\, A \boldsymbol{y}(t)\, dt\,. \tag{5.7}$$

*Die Konvergenz ist für $x > x_0$ fast gleichmäßig (d. h. gleichmäßig auf kompakten Intervallen) und für $x > \mu x_0$ außerdem absolut.*

Anmerkung: Man beachte, daß die $\boldsymbol{c}_n$ konstante Vektoren sind, so daß jede Lösung die (endliche oder unendliche) Summe von Linearkombinationen der Spaltenvektoren der Matrizen $F_n(x)$ ist. (Vgl. auch Satz 7).

Beweis: Zunächst folgt aus (4.5) und (5.2) formal unmittelbar (5.6), so daß nur die Zulässigkeit der Vertauschung von Integration und Summation zu zeigen ist. Wegen $\mu t \in [x_0, \mu x_0]$, (5.4) und (5.5) konvergiert die Reihe $\Sigma F_n(x)\, G_n(\mu t)$ für jedes feste $x > \mu x_0$ (im Sinne der Anmerkung nach Satz 5) gleichmäßig und absolut für $t \in [\lambda x_0, x_0]$. Für $x \in\, ]x_0, \mu x_0]$ ist wegen $\varphi_{n;jk}(x, \mu t) \equiv 0$ für $j \neq k$, nur das Verhalten von $\Sigma \varphi_{n;jj}(x, \mu t) = \Sigma f_n(x)\, g_n(\mu t)$, d. h. der skalare Fall (vgl. die Anmerkung nach Satz 2) zu untersuchen. Die Gültigkeit von (5.6) kann daher analog wie in [1], 6.5, gezeigt werden. Die Konvergenz bezüglich $x$ ist weiterhin fast gleichmäßig, sie muß jedoch nicht mehr absolut sein.

Anmerkung: Die Lösungen der adjungierten Gleichung besitzen ebenfalls eine Darstellung der Form (5.6). Ist $\boldsymbol{g}(x)$ für $x \leqq x_1$ Lösung der adjungierten Gleichung (2.1), so gilt (vgl. (4.6))

$$\boldsymbol{g}(x) = \sum_{n=-\infty}^{\infty} \langle \boldsymbol{g}, F_n\rangle\, G_n(x)\,, \qquad x < x_1\,. \tag{5.8}$$

Für $x < \lambda x_1$ ist die Konvergenz wieder absolut.

### 6. Das asymptotische Verhalten beliebiger Lösungen

Mit Hilfe der Darstellung (5.6), sowie den asymptotischen Formeln aus Punkt 3 kann nun das asymptotische Verhalten beliebiger Lösungen des AWP (3), (5) angegeben werden. Zuvor das folgende

Lemma 2: *Ist $\boldsymbol{y}(x)$ für $x \geqq x_0$ eine Lösung des* AWP (3), (5), *dann gilt für die Komponenten $c_{nk}$, $k = 1, 2, \dots, r$ der Vektoren $\boldsymbol{c}_n = \langle G_n, y\rangle$*

$$c_{nk} = O\left(\exp\left\{\frac{2\pi^2n^2}{c} - \frac{\pi^2|n|}{c} - \frac{\ln^2|n|}{c} - \frac{2\pi i n\gamma}{c} + b\ln|n|\right\}\right) \tag{6.1}$$

*mit einer geeignet gewählten Konstanten $b > 0$ und $\gamma = c/2 + \ln \omega c$.*

Beweis: Setzt man $\boldsymbol{z} = A\boldsymbol{y}$, so folgt aus (2.23)

$$c_{nk} = \sum_{j=k}^{r} q_{j-k}(\mu x_0)\, g_n(\mu^{j-k}x_0)\, y_j(x_0) + \mu \sum_{j=k}^{r} \int_{\lambda x_0}^{x_0} q_{j-k}(\mu^2 t)\, g_n(\mu^{j-k+1}t)\, z_j(t)\, dt\,.$$

Offenbar genügt es, das Verhalten der Summen $\sum_{j-k}^{r} g_n(\mu^{j-k}x)$, $x \in [x_0, \mu x_0]$ für $|n| \to \infty$ abzuschätzen. Mit (3.13) erhält man

$$g_n(\mu^{j-k}x) = O\left(g_n(1)\exp\left\{[\ln x + c(j-k)]\frac{\ln|n|}{c}\right\}\right), \tag{6.2}$$

und mit (3.6), (3.9) (vgl. auch (2.25))

$$g_n(1) = O\left(\exp\left\{\frac{2\pi^2n^2}{c} - \frac{\pi^2|n|}{c} - \frac{\ln^2|n|}{2c} - \frac{2\pi i n\gamma}{c} + \frac{\ln\omega c \ln|n|}{c}\right\}\right). \tag{6.3}$$

Aus (6.2), (6.3) folgt nun unmittelbar die Behauptung.

Der folgende Satz charakterisiert das asymptotische Verhalten beliebiger Lösungen.

Satz 7: *Ist $\boldsymbol{y}(x)$ eine Lösung des* AWP (3), (5) *und sind die $\pi_k(t)$, $k = 1, 2, \dots, r$, die c-periodischen Funktionen*

$$\pi_k(t) = \sum_{n=-\infty}^{\infty} c_{nk}\exp\left\{-\frac{2\pi^2n^2}{c} + \frac{2\pi i n\gamma}{c} + \frac{2\pi i n t}{c}\right\}, \tag{6.4}$$

*dann hat $\boldsymbol{y}(x)$ für $x \to \infty$ die asymptotische Form*

$$\boldsymbol{y}(x) = \sum_{k=1}^{r} \boldsymbol{y}_k(x) \tag{6.5}$$

*mit*

$$\boldsymbol{y}_k(x) = e^{h(\ln x)}\left[\pi_k\left(\ln x - \ln\ln x + \frac{\ln\ln x}{\ln x}\right) + O\left(\frac{1}{\ln x}\right)\right]\boldsymbol{v}_k(x)\,. \tag{6.6}$$

*Dabei ist der Vektor $\boldsymbol{v}_k(x)$ durch*

$$v_{kj}(x) = \begin{cases} \dfrac{(\omega c)^{-(k-j)}}{(k-j)!}\ln^{k-j} x\,, & j = 1, 2, \dots, k\,,\\ 0, & j = k+1, \dots, r\,,\end{cases} \tag{6.7}$$

*gegeben.*

Beweis:

1. (6.5): Wegen Satz 6 ist nur die Konvergenz der Reihen $\boldsymbol{y}_k(x) = \sum_{n=-\infty}^{\infty} c_{nk}\boldsymbol{f}_{nk}(x)$ für hinreichend große $x$ zu zeigen, wenn mit $\boldsymbol{f}_{nk}(x)$ die $k$-ten Spaltenvektoren der Matrizen $F_n(x)$ bezeichnet werden. Mit (2.23) erhält man

$$y_{kj}(x) = \sum_{n=-\infty}^{\infty} c_{nk}p_{k-j}(x)\, f_n(\lambda^{k-j}x)\,, \qquad j \leqq k\,. \tag{6.8}$$

Sei nun $x \in \mathbb{R}_+$ fest gewählt, so folgt aus (3.12), (3.14), (6.1) und (6.3) für $|n| \to \infty$

$$c_{nk}f_n(\lambda^{k-j}x) = O\left(\exp\left\{\left(\frac{1}{c}\ln\frac{x_0}{x} + r - j\right)\ln|n|\right\}\right),$$

d. h. die Reihe in (6.8) konvergiert gleichmäßig und absolut für $x \geqq a > \mu^{r-j}x_0$.

2. (6.6): Sind die $\boldsymbol{y}_k(x)$ nur endliche Summen, so folgt die in (6.6) angegebene asymptotische Form unmittelbar aus (3.5), (3.7) und (3.8), so daß nur der allgemeine Fall zu untersuchen ist. Dazu werden gleichmäßige Abschätzungen in $x$ und $n$ benötigt, die ähnlich wie im skalaren Fall gefunden werden können (vgl. auch [1], 8.1). Zunächst erhält man für hinreichend große $x$ und $|n| \geqq 2$ mit der Darstellung (3.3) für $h(\ln x)$ unter Verwendung von

$$\ln\left|\frac{c_{nk}p_{k-j}(x)\, f_n(\lambda^{k-j}x)}{f_0(x)}\right| = \ln|c_{nk}p_{k-j}(x)| + \ln\left|\frac{f_n(\lambda^{k-j}x)}{f_0(x)}\right|$$

die folgenden oberen Schranken. Die $b_i$ sind dabei geeignet gewählte positive Konstante.

$$\left|\frac{c_{nk} p_{k-j}(x)\, f_n(\lambda^{k-j}x)}{f_0(x)}\right| < \begin{cases} |v_{kj}(x)|\, e^{-b_1|n|}\,, & |n| < \ln x \\ |v_{kj}(x)|\, x^{-b_2}\,, & \ln x \leq |n| < \ln^2 x\,. \\ |v_{kj}(x)|\, e^{-b_3 \ln x \ln|n|}\,, & |n| \geq \ln^2 x \end{cases} \tag{6.9}$$

Für $|n| \leq \ln^\beta x$, $\beta < \frac{1}{2}$, ergibt sich mit (3.9), (6.1) und (6.4)

$$\left|c_{nk}\left\{\frac{p_{k-j}(x)\, f_n(\lambda^{k-j}x)}{v_{kj}(x)\, f_0(x)} - \exp\left[-\frac{2\pi^2 n^2}{c} + \frac{2\pi i n\gamma}{c} + \frac{2\pi i n}{c}\left(\ln x - \ln\ln x + \frac{\ln\ln x}{\ln x}\right)\right]\right\}\right| < \frac{e^{-b_4|n|}}{\ln x}. \tag{6.10}$$

Aus (6.1) und (6.4) folgt außerdem

$$\left|c_{nk} \exp\left\{-\frac{2\pi^2 n^2}{c} + \frac{2\pi i n\gamma}{c} + \frac{2\pi i n t}{c}\right\}\right| < b_5\, e^{-b_6|n|}\,, \qquad t \in \mathbb{R}\,, \tag{6.11}$$

sodaß man mit (6.9), (6.10) und (6.11) durch Summation über $n$

$$\left|\frac{e^{-h(\ln x)} y_{kj}(x)}{v_{kj}(x)} - \pi_k\left(\ln x - \ln\ln x + \frac{\ln\ln x}{\ln x}\right)\right| <$$

$$< \sum_{|n| \leq \ln^\beta x} \frac{e^{-b_4|n|}}{\ln x} + \sum_{\ln^\beta x < |n| < \ln x} e^{-b_1|n|} + \sum_{\ln x \leq |n| < \ln^2 x} x^{-b_2} + \sum_{|n| \geq \ln^2 x} |n|^{-b_3 \ln x} + b_5 \sum_{|n| \geq \ln^\beta x} e^{-b_6|n|}$$

erhält. Für $x \to \infty$ ist jede Summe ein $O\left(\frac{1}{\ln x}\right)$ sodaß (6.6) bewiesen ist.

Aus dem Beweis sieht man ohne Schwierigkeiten, daß auch die Umkehrung von Satz 7 gültig ist, d. h. sind die $\pi_k(t)$ vorgegebene $c$-periodische Funktionen, deren Koeffizienten in der Darstellung (6.4) der Beziehung (6.1) genügen, dann existiert eine Lösung $\boldsymbol{y}(x)$ mit dem asymptotischen Verhalten (6.5), (6.6). Durch (6.5), (6.6) wird außerdem eine lineare Abbildung $P$ von den Lösungen des AWP (3), (5) in die $c$-periodischen (Vektor) Funktionen $\boldsymbol{\pi}(t)$ definiert. Diese Abbildung ist offenbar injektiv, denn aus $P\boldsymbol{y} = 0$ folgt mit (6.4) $c_{nk} = 0$, $n \in \mathbb{Z}$, $k = 1, 2, \ldots, r$, und daher mit (5.6) $\boldsymbol{y}(x) \equiv 0$.

Satz 7 kann daher zu folgendem Existenz- und Eindeutigkeitssatz verschärft werden.

Satz 8: *Sind die* $\pi_k(t)$, $k = 1, 2, \ldots, r$, *gegebene c-periodische Funktionen, deren Koeffizienten* $c_{nk}$ *bezüglich der Darstellung* (6.4) *die asymptotische Form* (6.1) *besitzen, dann existiert für geeignet gewählte* $x_0 \in \mathbb{R}_+$ *genau ein* $\boldsymbol{\alpha}(x) \in C(I_0)$, *sodaß die Lösung des AWP* (3), (5) *das asymptotische Verhalten* (6.5), (6.6) *besitzt.*

Anmerkung: Alle bisher vorliegenden Ergebnisse wurden für $\omega \neq 0$ bewiesen. Für $\omega = 0$ hat das AWP (3), (5) bei beliebig vorgegebener Anfangsfunktion $\boldsymbol{\alpha}(x)$ i. a. keine Lösung mit den Eigenschaften (6) bis (8), während es bei geeigneter Vorgabe von $\boldsymbol{\alpha}(x)$ nur Polynomlösungen besitzt (vgl. auch Satz 1). Der Grad des in der ersten Komponente der Lösung auftretenden Polynoms ist offenbar höchstens $r - 1$. Die Verhältnisse für Systemmatrizen $A$ in beliebiger Form ergeben sich durch die Transformation $\boldsymbol{y}(x) = T\boldsymbol{u}(x)$, wenn $T$ eine Matrix ist, mit deren Hilfe $A$ auf NF gebracht werden kann.

## Literatur


1 DE BRUIJN, N. G., The difference-differential equation $F'(x) = e^{\alpha x + \beta} F(x - 1)$. I, II, Indagationes Math. 15, 449—464 (1953).
2 CARR, J.; DYSON, J., The matrix functional-differential equation $y'(x) = Ay(\lambda x) + By(x)$, Lecture Notes in Mathematics, vol. 415, pp. 323—328, Berlin-Heidelberg-New York, Springer 1974.
3 ENG-BIN LIM, Asymptotic behavior of solutions of the functional-differential equation $x'(t) = Ax(\lambda t) + Bx(t)$, J. Math. Analysis Appl. 55, 794—806 (1976).
4 HAHN, W., Über die Funktional-Differentialgleichung $f'(z) = f(qz)$ und verwandte Funktionalgleichungen, Anm. Univ. Sci. Budapest. Eötvös Sect. Math., 3—21 (1973).
5 KATO, T.; MCLEOD, J. B., The functional-differential equation $y'(x) = ay(\lambda x) + by(x)$, Bull. Amer. Math. Soc. 77, 891—937 (1971).
6 MCLEOD, J. B., The functional-differential equation $y'(x) = ay(\lambda x) + by(x)$ and generalisations, Lecture Notes in Mathematics, vol. 280, pp. 308—313, Berlin-Heidelberg-New York, Springer 1972
7 OCKENDON, J. R.; TAYLER, A. B., The dynamics of a current collection system for an electric locomotive, Proc. Roy. Soc. London, Ser. A 322, 447—468 (1971).


Eingereicht: 26. 11. 1978

*Anschrift*: Dipl.-Ing. Dr. FRITZ VOGL, Institut für Technische Mathematik an der TU Wien, Gußhausstraße 27—29, A-1040 Wien, Österreich

ZAMM 60, 19 – 24 (1980)

K. R. SCHNEIDER

# Andronov-Hopf-Bifurkation im Falle stetiger Parameterabhängigkeit

*Wir betrachten das Problem der Abzweigung einer periodischen Lösung von einer Gleichgewichtslösung eines zweidimensionalen autonomen Differentialgleichungssystems* $dx/dt = f(x, \lambda)$, *falls* $\lambda$ *einen kritischen Wert* $\lambda_0$ *durchläuft. Unter den Voraussetzungen, daß* $f$ *bezüglich* $x$ *dreimal stetig differenzierbar ist, von* $\lambda$ *aber nur stetig abhängt und daß beim Durchgang von* $\lambda$ *durch* $\lambda_0$ *nicht mehr als ein Grenzzyklus von der betrachteten stationären Lösung abzweigt (generischer Fall, der durch ein Koeffizientenkriterium charakterisiert wird), leiten wir ein Kriterium über die Bifurkationsrichtung ab, das diejenigen Kriterien verallgemeinert, die auf der Differenzierbarkeit von* $f$ *bezüglich* $\lambda$ *basieren. Das Resultat ist auch auf* $n$*-dimensionale Systeme anwendbar, die sich mittels einer Zentrumsmannigfaltigkeit auf zweidimensionale Systeme reduzieren lassen.*

*Consider the problem of bifurcation of a periodic solution from an equilibrium point of a two-dimensional autonomous differential system* $dx/dt = f(x, \lambda)$ *where* $\lambda$ *passes through a critical value* $\lambda_0$. *Suppose that* $f$ *is continuously differentiable with respect to* $x$ *up to the order three but depends on* $\lambda$ *only continuously, and that at most one limit cycle bifurcates from the considered stationary solution when* $\lambda$ *crosses* $\lambda_0$ *(generic case, it will be characterized by a coefficients criterion). We deduce a criterion on the bifurcation direction generalizing those criteria being based on the differentiability of* $f$ *with respect to* $\lambda$. *The result can be applied to* $n$*-dimensional systems which can be reduced to a two-dimensional system using a center manifold.*

Рассмотрим задачу бифуркации периодического решения из точки равновесия двумерной автономной системы дифференциальных уравнений $dx/dt = f(x, \lambda)$, когда $\lambda$ проходит через критическое значение $\lambda_0$. Предположим, что $f$ непрерывно дифференцируема по $x$ до третьего порядка, но зависит от $\lambda$ только непрерывно, и что порождается не больше одного предельного цикла из рассмотренного стационарного решения при прохождении параметра $\lambda$ через $\lambda_0$ (общий случай характеризован критерием коэффициентов). Выводим критерий о направлении бифуркации, обобщающий те критерии, которые основаны на дифференцируемости функции $f$ по $\lambda$. Полученный результат можно тоже применить к $n$-мерным системам, допускающие сведение к двумерным системам.

**1.** Bei der mathematischen Modellierung zahlreicher biochemischer, biologischer und ökologischer Prozesse spielen Systeme gewöhnlicher autonomer Differentialgleichungen, die von gewissen Parametern abhängen, eine wichtige Rolle [3, 6, 12, 13, 16, 17]. Die Bestimmung des Verhaltens solcher Modelle erfordert eine Untersuchung der topologischen Struktur der Integralkurven des zugehörigen dynamischen Systems in Abhängigkeit von den auftretenden Parametern. Eine der Möglichkeiten für die Änderung des qualitativen Verhaltens autonomer Systeme besteht in der Abzweigung periodischer Lösungen von stationären Lösungen, falls ein gewisser Parameter $\lambda$ einen kritischen Wert $\bar{\lambda}$ durchläuft. Wir nehmen an, daß das $n$-dimensionale dynamische System

$$\frac{dx}{dt} = f(x, \lambda)\,, \qquad x \in R^n,\ \lambda \in R\,, \tag{1.1}$$

für $\lambda = \bar{\lambda}$ einen singulären Punkt $\bar{x}$ besitzt, d. h. $f(\bar{x}, \bar{\lambda}) = 0$, so daß folgende Eigenschaften gelten:

1. $f(x, \lambda)$ ist in einer Umgebung von $(\bar{x}, \bar{\lambda})$ nach $x$ differenzierbar, wobei $f_x(x, \lambda)$ stetig von $(x, \lambda)$ abhängt.

2. Die Eigenwerte der JACOBI-Matrix $f_x(\bar{x}, \bar{\lambda})$ sind sämtlich von Null verschieden, unter ihnen gibt es genau ein Paar (algebraisch einfache) rein imaginäre Eigenwerte.

In diesem Fall kann das Problem der Abzweigung einer periodischen Lösung vom singulären Punkt $\bar{x}(\bar{\lambda})$ unter Verwendung der sogenannten Zentrumsmannigfaltigkeit [7, 9, 14] auf ein analoges Verzweigungsproblem eines zweidimensionalen autonomen Systems zurückgeführt werden. Im folgenden beschränken wir uns deshalb auf das Problem der Bifurkation einer isolierten periodischen Lösung (Grenzzyklus) aus dem singulären Punkt eines zweidimensionalen autonomen Systems, das von einem Parameter abhängt. Die dabei auftretenden Probleme können wie folgt charakterisiert werden:

1. Notwendige und hinreichende Bedingungen für die Abzweigung eines Grenzzyklus von einem singulären Punkt.
2. Anzahl der abzweigenden Grenzzyklen.
3. Stabilitätsverhalten der Grenzzyklen.
4. Bifurkationsrichtung.
5. Asymptotisches Verhalten der periodischen Lösungen.

Bereits am Ende des vorigen Jahrhunderts beschäftigte sich H. POINCARÉ [10] mit einigen der genannten Fragestellungen. Aber erst ein halbes Jahrhundert später steht diese Problematik wieder im Mittelpunkt zahlreicher Untersuchungen. Angeregt durch wichtige praktische Aufgaben begannen A. A. ANDRONOV und seine Mitarbeiter in der Stadt Gorki mit der Entwicklung der Bifurkationstheorie ebener dynamischer Systeme (eine diesbezügliche Literaturübersicht findet man in [1]). Hinsichtlich der Bifurkation von Grenzzyklen aus einem singulären Punkt vom Fokustyp, d. h. einem singulären Punkt, dessen charakteristische Wurzeln rein imaginär sind, kann unter Verwendung der Resultate von A. A. ANDRONOV und seiner Mitarbeiter [2, § 25.3] folgender Satz formuliert werden:

Satz 1.1: *Es sei $G \subset R^2$ eine Umgebung des Punktes $x = 0$, $\Lambda$ sei das Intervall $\Lambda = \{\lambda \in R^1: |\lambda| < \lambda_0, \lambda_0 > 0\}$. Unter den Voraussetzungen*

$A_1$. *$f: G \times \Lambda \to R^2$ ist bezüglich $x$ dreimal und bezüglich $\lambda$ einmal stetig differenzierbar;*

$A_2$. *$f(0, \lambda) \equiv 0$ für alle $\lambda \in \Lambda$;*

$A_3$. *$f_x(0, \lambda)$ besitzt für $\lambda \in \Lambda$ konjugiert komplexe Eigenwerte $\alpha(\lambda) \pm i\beta(\lambda)$, die den Relationen $\alpha(0) = 0$, $\alpha'(0) \neq 0$, $\beta(0) \neq 0$ genügen;*

$A_4$. *die Fokusgröße $\alpha_3(0)$* [Definition im folgenden Abschnitt] *des singulären Punktes $x = 0$ des Systems (1.1) für $\lambda = 0$ genügt der Relation $\alpha_3(0) \neq 0$;*

*zweigt beim Durchgang von $\lambda$ durch 0 genau ein Grenzzyklus $\Gamma(\lambda)$ vom singulären Punkt $x = 0$ ab, wobei die Bifurkationsrichtung durch* sgn $[-\alpha'(0)\,\alpha_3(0)]$ *bestimmt wird, d. h., das Vorzeichen von $-\alpha'(0)\,\alpha_3(0)$ gibt an, ob für (hinreichend kleine) positive oder negative $\lambda$ in einer hinreichend kleinen Umgebung von $x = 0$ genau ein Grenzzyklus $\Gamma(\lambda)$ von (1.1) existiert. Das Stabilitätsverhalten des Grenzzyklus $\Gamma(\lambda)$ ist entgegengesetzt zu dem des singulären Punktes $x = 0$ des Systems (1.1).*

1967 wurde vom Verfasser gezeigt [13, Satz 22], daß die Bedingungen $\alpha(0) = 0$, $\alpha'(0) \neq 0$ in Satz 1.1 unter der Voraussetzung, daß $\alpha(\lambda)$ im Punkt 0 hinreichend oft differenzierbar ist, durch $\alpha(0) = 0, \alpha'(0) = 0, \ldots, \alpha^{(k)}(0) \neq 0$ ersetzt werden können, wobei die Funktion $\alpha(\lambda)$ im Punkt $\lambda = 0$ ihr Vorzeichen wechsle. Die Bifurkationsrichtung wird dann durch sgn $[\alpha(-\lambda)\,\alpha_3(0)]$, $\lambda$ hinreichend klein und positiv, bestimmt. Äquivalent zu dieser Aussage ist die folgende:

Corollar 1.1: *Unter der Voraussetzung, daß $\alpha(\lambda)$ im Punkt $\lambda = 0$ hinreichend oft differenzierbar ist, können die Bedingungen $\alpha(0) = 0$, $\alpha'(0) \neq 0$ in Satz 1.1 durch $\alpha(0) = \alpha'(0) = \ldots = \alpha^{(k-1)}(0) = 0$, $\alpha^{(k)}(0) \neq 0$, $k$ ungerade, ersetzt werden, wobei dann die Bifurkationsrichtung durch* sgn $[-\alpha^{(k)}(0)\,\alpha_3(0)]$ *bestimmt wird.*

Ein ähnliches Resultat wurde 1977 von D. S. Schmidt [9] publiziert, dessen Beweis auf einer Parametrisierung von $\lambda$ und der Anwendung des Satzes über implizite Funktionen auf eine zugehörige Bifurkationsgleichung beruht.

1942 wurde das Problem der Abzweigung einer periodischen Lösung von einem singulären Punkt eines $n$-dimensionalen Systems (1.1) von E. Hopf [5] untersucht. Für den zweidimensionalen Fall kann sein Resultat unter Berücksichtigung, daß die von ihm geforderte Analytizität nicht notwendig ist [9, S. 197], wie folgt formuliert werden.

Satz 1.2: *Unter den Voraussetzungen $A_1$ bis $A_3$ existiert für hinreichend kleines $\varepsilon$ eine Familie periodischer Lösungen $x = p(t, \varepsilon) \not\equiv 0$, $\lambda = \lambda(\varepsilon)$ mit $p(t, 0) \equiv 0$, deren Periode $\tau(\varepsilon)$ für $\varepsilon \to 0$ beschränkt bleibt.*

Bemerkung 1.2.1: Nach den Resultaten in [15] gilt unter der Voraussetzung $A_3$ stets $\tau(0) < +\infty$.

Bemerkung 1.2.2: Ist $f(x, \lambda)$ bezüglich $x$ analytisch, dann zweigt unter der Voraussetzung $\alpha'(0) \neq 0$ beim Durchgang von $\lambda$ durch 0 kein Grenzzyklus ab, falls $x = 0$ für $\lambda = 0$ ein Zentrumspunkt ist [8].

Bemerkung 1.2.3: E. Hopf [5] leitet keine Formel für die Bifurkationsrichtung ab. Die ersten Formeln für die Bestimmung der Bifurkationsrichtung stammen von I. D. Hsü und N. D. Kazarinoff [6] und A. B. Poore [11], wobei in beiden Fällen die Differenzierbarkeit von $f(x, \lambda)$ nach $\lambda$ vorausgesetzt wird.

Bemerkung 1.2.4: Es lassen sich leicht zweidimensionale Systeme (1.1) angeben, für die die Relation $\alpha'(0) = 0$ gilt und bei denen zwei geschlossene Lösungskurven vom singulären Punkt $x = 0$ abzweigen. Aus Corollar 1.1 folgt aber, daß $\alpha'(0) \neq 0$ keine notwendige Bedingung für die Abzweigung von genau einem Grenzzyklus aus $x = 0$ darstellt.

Im folgenden wollen wir das Problem der Bifurkation einer periodischen Lösung aus dem singulären Punkt $x = 0$ eines ebenen dynamischen Systems (1.1) untersuchen, wobei wir $\alpha_3(0) \neq 0$ voraussetzen, von $f(x, \lambda)$ aber nur eine stetige Abhängigkeit bezüglich $\lambda$ fordern.

1. Wir betrachten das zweidimensionale Differentialgleichungssystem

$$\frac{dx}{dt} = f(x, \lambda) \tag{2.1}$$

unter folgenden Voraussetzungen über $f$:

$V_1$. $f: G \times \Lambda \to R^2$ ist bezüglich $x$ dreimal differenzierbar, wobei alle Ableitungen in $G \times \Lambda$ stetig von $(x, \lambda)$ abhängen. $G$ stellt eine Umgebung von $x = 0$ dar, $\Lambda$ ist durch $\Lambda = \{\lambda \in R : |\lambda| < \lambda_0, \lambda_0 > 0\}$ definiert.

$V_2$. $f(0, \lambda) \equiv 0$ für alle $\lambda \in \Lambda$. (2.2)

$V_3$. $J(\lambda) := f_x(0, \lambda)$ besitzt für $\lambda \in \Lambda$ konjugiert komplexe Eigenwerte $\alpha(\lambda) \pm i\beta(\lambda)$, die den Relationen genügen

$$\beta(\lambda) > 0 \quad \text{für alle } \lambda \in \Lambda, \tag{2.3}$$

$$\alpha(0) = 0, \quad \alpha(\lambda) \neq 0 \quad \text{für } \lambda \neq 0 \text{ und } \lambda \in \Lambda. \tag{2.4}$$

Unter der Voraussetzung $V_3$ existiert für alle $\lambda \in \Lambda$ eine nicht singuläre Matrix $P(\lambda)$, so daß

$$P(\lambda)^{-1}\,J(\lambda)\,P(\lambda) = \begin{pmatrix} \alpha(\lambda) & -\beta(\lambda) \\ \beta(\lambda) & \alpha(\lambda) \end{pmatrix} \tag{2.5}$$

gilt. Unter Verwendung der Koordinatentransformation $y = P(\lambda) \cdot x$ können wir folglich das System (2.1) in der Gestalt

$$\frac{dy_1}{dt} = \alpha(\lambda) y_1 - \beta(\lambda) y_2 + g_1(y, \lambda), \quad \frac{dy_2}{dt} = \beta(\lambda) y_1 + \alpha(\lambda) y_2 + g_2(y, \lambda) \tag{2.6}$$

darstellen, wobei $g(y, \lambda)$ dieselben Regularitätseigenschaften wie $f(x, \lambda)$ besitzt und für $|y| = \sqrt{y_1^2 + y_2^2} \to 0$ gleichmäßig bezüglich $\lambda$, $|\lambda| \leqq \lambda_1$, $0 < \lambda_1 < \lambda_0$, der Relation $g(y, \lambda) = o(|y|)$ genügt.

Unter Verwendung von Polarkoordinaten $r, \varphi$ ($y_1 = r \cos \varphi$, $y_2 = r \sin \varphi$) ist (2.6) für $|\lambda| \leqq \lambda_1$ in einer hinreichend kleinen Umgebung von $y = 0$ ($0 \leqq r < r_0$, $0 \leqq \varphi \leqq 2\pi$) äquivalent zur Differentialgleichung

$$\frac{dr}{d\varphi} = \left\{ \frac{\alpha(\lambda)}{\beta(\lambda)} r + \frac{\tilde{g}_1(r, \varphi, \lambda) \cos \varphi + \tilde{g}_2(r, \varphi, \lambda) \sin \varphi}{\beta(\lambda)} \right\} \left\{ 1 - \frac{\tilde{g}_1(r, \varphi, \lambda) \sin \varphi - \tilde{g}_2(r, \varphi, \lambda) \cos \varphi}{\beta(\lambda) r} \right\}^{-1}, \tag{2.7}$$

die wir für $0 \leqq r \leqq r_1 \leqq r_0$ und $|\lambda| \leqq \lambda_1$ in der Gestalt

$$\frac{dr}{d\varphi} = k_1(\varphi, \lambda) r + k_2(\varphi, \lambda) r^2 + k_3(\varphi, \lambda) r^3 + r^3 k(r, \varphi, \lambda) \tag{2.8}$$

schreiben können. Über die Funktionen $k_i(\varphi, \lambda)$ und $k(r, \varphi, \lambda)$ gilt folgendes Lemma.

Lemma 2.1: *Unter den Voraussetzungen* $V_1$ *bis* $V_3$ *sind die Funktionen* $k_i(\varphi, \lambda)$, $i = 1, 2, 3$, *homogene Polynome in* $\cos \varphi$ *und* $\sin \varphi$, *deren Grad gerade (ungerade) ist, falls* $i$ *ungerade (gerade) ist. Dabei gilt*

$$k_1(\varphi, \lambda) = \frac{\alpha(\lambda)}{\beta(\lambda)}. \tag{2.9}$$

$k(r, \varphi, \lambda)$ *ist bezüglich* $\varphi$ *dreimal stetig differenzierbar und bezüglich* $\lambda$ *stetig, außerdem genügt* $k(r, \varphi, \lambda)$ *gleichmäßig bezüglich* $\varphi$ *und* $\lambda$, $0 \leqq \varphi \leqq 2\pi$, $|\lambda| \leqq \lambda_1$, *der Relation* $k(r, \varphi, \lambda) = o(1)$ *für* $r \to 0$.

Ein wichtiges Hilfsmittel für die Untersuchung des Verhaltens der Trajektorien der Gleichung (2.8) in der Umgebung des Punktes $r = 0$ ist die sogenanten POINCARÉsche Zuwachsfunktion. Zu ihrer Definition bezeichnen wir mit $h(\varphi; \varrho, \lambda)$ eine Lösung der Differentialgleichung (2.8), die der Anfangsbedingung

$$h(0; \varrho, \lambda) = \varrho, \quad 0 \leqq \varrho \leqq r_2 \leqq r_1 \tag{2.10}$$

genügt. Dabei soll $r_2$ so klein gewählt sein, daß $h(\varphi; \varrho, \lambda)$ im Intervall $0 \leqq \varphi \leqq 4\pi$ existiert.

Definition 2.1: Die Abbildung

$$\psi(\varrho, \lambda) := h(2\pi; \varrho, \lambda) - \varrho, \quad 0 \leqq \varrho \leqq r_2, \quad |\lambda| \leqq \lambda_1 \tag{2.11}$$

wird als POINCARÉ*sche Zuwachsfunktion* bezeichnet.

Bemerkung 2.1.1: Die POINCARÉsche Zuwachsfunktion stellt den Zuwachs der Trajektorie $h(\varphi; \varrho, \lambda)$ auf dem Strahl $\varphi = 0$ nach einem vollständigen Umlauf um den singulären Punkt $r = 0$ dar. Einer Nullstelle $\varrho(\lambda)$ der Funktion $\psi(\varrho, \lambda)$ entspricht somit eine geschlossene Lösungskurve $\Gamma(\lambda)$ von (2.8), die durch den Punkt $(\varrho(\lambda), 0)$ geht. $\Gamma(\lambda)$ ist dabei *orbital stabil* (*instabil*), falls $\psi(\varrho, \lambda)$ an der Stelle $\varrho = \varrho(\lambda)$ für wachsendes $\varrho$ das Vorzeichen von plus nach minus (von minus nach plus) ändert.

Bemerkung 2.1.2: Obwohl $\psi(\varrho, \lambda)$ in (2.11) nur für positive $\varrho$ definiert ist, kann $\psi(\varrho, \lambda)$ für hinreichend kleines $\varrho$ durch die geometrisch plausible Definition $\psi(-\varrho, \lambda) := -(h(3\pi; \varrho, \lambda) - h(\pi; \varrho, \lambda))$ auch für negatives $\varrho$ definiert werden [2].

Bemerkung 2.1.3: Unter Verwendung der Funktion $\psi(\varrho, \lambda)$ kann das Problem der Bifurkation periodischer Lösungen der Differentialgleichung (2.8) vom singulären Punkt $r = 0$ auf das Problem der Abzweigung von positiven Nullstellen der Funktion $\psi(\varrho, \lambda)$ von der trivialen Nullstelle $\varrho = 0$ zurückgeführt werden.

Unter der Voraussetzung $V_1$ ist $h(\varphi; \varrho, \lambda)$ dreimal stetig nach $\varrho$ differenzierbar. Beachten wir noch, daß $r \equiv 0$ eine Lösung von (2.8) ist, dann läßt sich $h(\varphi; \varrho, \lambda)$ in der Gestalt

$$h(\varphi; \varrho, \lambda) = \sum_{i=1}^{3} l_i(\varphi, \lambda) \varrho^i + \varrho^3 l(\varphi; \varrho, \lambda) \tag{2.12}$$

darstellen, wobei $l(\varphi; \varrho, \lambda)$ eine stetige Funktion ihrer Variablen ist und der Relation $l(\varphi; \varrho, \lambda) = o(1)$ für $\varrho \to 0$ gleichmäßig bezüglich $\varphi$ und $\lambda$, $0 \leqq \varphi \leqq 2\pi$, $|\lambda| \leqq \lambda_1$, genügt. Durch Einsetzen von (2.12) in (2.8) erhalten wir durch Koeffizientenvergleich zur Bestimmung der Funktionen $l_i(\varphi, \lambda)$ das Differentialgleichungssystem

$$\frac{dl_1}{dt} = l_1 k_1(\varphi, \lambda), \quad \frac{dl_2}{dt} = l_2 k_1(\varphi, \lambda) + l_1^2 k_2(\varphi, \lambda), \quad \frac{dl_3}{dt} = l_3 k_1(\varphi, \lambda) + 2 l_1 l_2 + l_1^3 k_3(\varphi, \lambda). \tag{2.13}$$

Dieses System bestimmt unter Berücksichtigung der aus (2.10) und (2.12) folgenden Anfangsbedingungen

$$l_1(0,\lambda) = 1\,, \qquad l_k(0,\lambda) = 0 \quad \text{für } k = 2, 3 \tag{2.14}$$

eindeutig die Funktionen $l_i(\varphi, \lambda)$, $i = 1, 2, 3$. Für $i = 1$ folgt unmittelbar nach (2.9) und (2.13)

$$l_1(\varphi,\lambda) = \exp\left[\frac{\alpha(\lambda)}{\beta(\lambda)}\varphi\right]. \tag{2.15}$$

Unter Verwendung von (2.12) erhalten wir nach (2.11) für $\psi(\varrho, \lambda)$ die Darstellung

$$\psi(\varrho,\lambda) = \sum_{i=1}^{3} \alpha_i(\lambda)\,\varrho^i + \varrho^3 l(2\pi;\varrho,\lambda)\,, \tag{2.16}$$

wobei die Koeffizienten $\alpha_i(\lambda)$ durch die Relationen

$$\alpha_1(\lambda) := l_1(2\pi,\lambda) - 1 = \exp\left[\frac{\alpha(\lambda)}{\beta(\lambda)}2\pi\right] - 1\,, \tag{2.17}$$

$$\alpha_k(\lambda) := l_k(2\pi,\lambda)\,, \qquad k = 2, 3\,, \tag{2.18}$$

definiert sind. Unter unseren Voraussetzungen sind die Koeffizienten $\alpha_i(\lambda)$, $i = 1, 2, 3$, für $|\lambda| \leqq \lambda_1$ stetige Funktionen von $\lambda$.

Definition 2.2: Die Zahlen $\alpha_i(0)$, $i = 1, 2, 3$, werden als *i-te Fokusgrößen* des singulären Punktes $x = 0$ des Systems (2.1) für $\lambda = 0$ bezeichnet.

Lemma 2.2: *Unter den Voraussetzungen* $V_1$ *bis* $V_3$ *gelten für* $|\lambda| \leqq \lambda_1$ *die Relationen*

$$\alpha_1(\lambda) = \alpha(\lambda)\,\gamma_1(\lambda)\,, \qquad \alpha_2(\lambda) = \alpha(\lambda)\,\gamma_2(\lambda)\,, \tag{2.19), (2.20}$$

*wobei* $\gamma_1(\lambda)$ *und* $\gamma_2(\lambda)$ *für* $|\lambda| \leqq \lambda_1$ *stetige Funktionen sind und* $\gamma_1(0)$ *der Beziehung*

$$\gamma_1(0) = \frac{2\pi}{\beta(0)} > 0 \tag{2.21}$$

*genügt.*

Beweis: Die Relationen (2.19) und (2.21) folgen unmittelbar aus (2.17) und (2.3). Die Integration der zweiten Gleichung von (2.13) ergibt unter Berücksichtigung von (2.9), (2.14) und (2.15)

$$l_2(\varphi,\lambda) = \exp\left[\frac{\alpha(\lambda)}{\beta(\lambda)}\varphi\right]\int_0^{\varphi} \exp\left[\frac{\alpha(\lambda)}{\beta(\lambda)}\tau\right] k_2(\tau,\lambda)\,\mathrm{d}\tau\,. \tag{2.22}$$

Setzen wir $\varphi = 2\pi$, so erhalten wir durch partielle Integration und unter Berücksichtigung von Lemma 2.1

$$\alpha_2(\lambda) = -\frac{\alpha(\lambda)}{\beta(\lambda)}\int_0^{2\pi} \exp\left[\frac{\alpha(\lambda)}{\beta(\lambda)}(2\pi + \tau)\right]\int_0^{\tau} k_2(\sigma,\lambda)\,\mathrm{d}\sigma\,\mathrm{d}\tau \equiv \alpha(\lambda)\,\gamma_2(\lambda)\,,$$

wobei $\gamma_2(\lambda)$ für $|\lambda| \leqq \lambda_1$ stetig ist.

Definition 2.3: Der singuläre Punkt $x = 0$ des Systems (2.1) wird für $\lambda = 0$ als *strukturell instabiler Fokus der Vielfachheit* 1 bezeichnet, falls die Relationen

$$\alpha(0) = 0\,, \qquad \alpha_3(0) \neq 0 \tag{2.23}$$

gelten.

Bemerkung 2.3.1: Der Terminus „strukturell instabil" deutet auf die Möglichkeit der Änderung der qualitativen Struktur der Integralkurven des Systems (2.1) in der Umgebung des singulären Punktes $x = 0$ hin, falls $\lambda$ den Wert $\lambda = 0$ durchläuft.

Bemerkung 2.3.2: Unter den Voraussetzungen $V_1$ bis $V_3$ ist ein strukturell instabiler Fokus der Vielfachheit 1 orbital stabil (orbital instabil), falls $\alpha_3(0)$ negativ (positiv) ist. Dies folgt unmittelbar aus (2.16) und Bemerkung 2.1.1.

Bemerkung 2.3.3: Ein notwendiges und hinreichendes Kriterium dafür, daß der singuläre Punkt $z = 0$ des Systems

$$\begin{aligned}\frac{\mathrm{d}z_1}{\mathrm{d}t} &= \sum_{1\leq i+j\leq 3} a_{ij}^{(1)}(\lambda)\,z_1^i z_2^j + r_1(z_1, z_2, \lambda)\,,\\ \frac{\mathrm{d}z_2}{\mathrm{d}t} &= \sum_{1\leq i+j\leq 3} a_{ij}^{(2)}(\lambda)\,z_1^i z_2^j + r_2(z_1, z_2, \lambda)\end{aligned} \tag{2.24}$$

einen strukturell instabilen Fokus der Vielfachheit 1 darstellt, wobei die Funktion $r(z, \lambda)$ dieselben Regularitätseigenschaften wie $f(x, \lambda)$ besitze und für $|\lambda| \leqq \lambda^*$ gleichmäßig bezüglich $\lambda$ der Relation $r(z, \lambda) = o(|z|^3)$ genüge, ist die Gültigkeit der Relationen [2, 12]

$$a_{10}^{(1)}(0) = a_{01}^{(2)}(0) = 0\,, \qquad a_{01}^{(1)}(0) = -a_{10}^{(2)}(0) = -\beta(0) \neq 0\,, \tag{2.25}$$

$$\alpha_3(0) = \frac{\pi}{4\beta(0)} \left\{ \frac{1}{\beta(0)} [a_{11}^{(1)}(a_{20}^{(1)} + a_{02}^{(1)}) + 2(a_{02}^{(1)}a_{02}^{(2)} - a_{20}^{(1)}a_{20}^{(2)}) - a_{11}^{(2)}(a_{20}^{(2)} + a_{02}^{(2)})] + 3(a_{30}^{(1)} + a_{03}^{(2)}) + a_{12}^{(1)} + a_{21}^{(2)} \right\}_{\lambda=0} \neq 0\,. \tag{2.26}$$

Im folgenden untersuchen wir das Problem der Bifurkation periodischer Lösungen des Systems (2.1) vom singulären Punkt $x = 0$ im generischen Fall, d. h. wir treffen die Voraussetzung:

$V_4$. Der singuläre Punkt $x = 0$ des Systems (2.1) stellt für $\lambda = 0$ einen strukturell instabilen Fokus der Vielfachheit 1 dar.

Nach (2.16) und Lemma 2.2 können wir die Poincarésche Zuwachsfunktion in der Gestalt

$$\psi(\varrho, \lambda) = \varrho\{\alpha(\lambda)\,[\gamma_1(\lambda) + \gamma_2(\lambda)\,\varrho] + [\alpha_3(\lambda) + l(2\pi; \varrho, \lambda)]\,\varrho^2\}$$

schreiben. Setzen wir

$$\tilde{\psi}(\varrho, \lambda) := \alpha(\lambda)\,[\gamma_1(\lambda) + \gamma_2(\lambda)\,\varrho] + [\alpha_3(\lambda) + l(2\pi; \varrho, \lambda)]\,\varrho^2\,, \tag{2.27}$$

so besitzt $\tilde{\psi}(\varrho, 0)$ unter den Voraussetzungen $V_1$ bis $V_4$ im Punkt $\varrho = 0$ eine Nullstelle zweiter Ordnung. Über die Aufspaltung dieser Nullstelle bei Variation von $\lambda$ gilt folgendes Lemma.

Lemma 2.3: *Unter den Voraussetzungen $V_1$ bis $V_4$ existieren positive Zahlen $\lambda_3$ und $\varrho_3$, so daß für alle $\lambda$ aus dem Intervall $|\lambda| \leqq \lambda_3$, die der Bedingung $\alpha_3(0)\,\alpha(\lambda) < 0$ genügen, folgende Aussagen gelten:*

1. *$\tilde{\psi}(\varrho, \lambda)$ besitzt im Intervall $|\varrho| \leqq \varrho_3$ genau zwei Nullstellen $\varrho_1(\lambda)$ und $\varrho_2(\lambda)$.*
2. *Diese Nullstellen sind algebraisch einfach und besitzen verschiedenes Vorzeichen.*

Beweis: Unter den getroffenen Voraussetzungen sind $\alpha_3(\lambda)$ und $l(2\pi; \varrho, \lambda)$ für $|\lambda| \leqq \lambda_1$ und $|\varrho| \leqq \varrho_1$ stetige Funktionen ihrer Variablen. Wegen $\alpha_3(0) \neq 0$ und $l(2\pi; \varrho, \lambda) = o(1)$ für $\varrho \to 0$ gleichmäßig bezüglich $\lambda$, $|\lambda| \leqq \lambda_1$, existieren positive Zahlen $\varrho_2$ und $\lambda_2$, $\varrho_2 \leqq \varrho_1$, $\lambda_2 \leqq \lambda_1$, so daß für $|\varrho| \leqq \varrho_2$ und $|\lambda| \leqq \lambda_2$ die Relationen

$$|\alpha_3(\lambda) + l(2\pi; \varrho, \lambda)| \geqq c_1 > 0\,, \tag{2.28}$$

$$\left|\frac{\partial^2\tilde{\psi}(\varrho, \lambda)}{\partial\varrho^2}\right| \geqq c_2 > 0$$

gelten, so daß nach dem Satz von Rolle $\tilde{\psi}(\varrho, \lambda)$ für $|\lambda| \leqq \lambda_2$ im Intervall $|\varrho| \leqq \varrho_2$ höchstens zwei Nullstellen besitzt. Unter Verwendung der Funktion

$$g(\varrho, \lambda) := + \left\{\frac{-\alpha(\lambda)}{\alpha_3(\lambda) + l(2\pi;\, \varrho, \lambda)} [\gamma_1(\lambda) + \gamma_2(\lambda)\,\varrho]\right\}^{1/2} \tag{2.29}$$

ist die Gleichung $\psi(\varrho, \lambda) = 0$ für $|\lambda| \leqq \lambda_2$ und $|\varrho| \leqq \varrho_2$ nach (2.27) und (2.28) äquivalent zu den Gleichungen

$$\varrho = g(\varrho, \lambda)\,, \qquad \varrho = -g(\varrho, \lambda)\,. \tag{2.30}$$

Damit diese Gleichungen reelle Lösungen besitzen, darf der Radikand in (2.29) nicht negativ werden. Nach (2.4) und (2.28) gilt für $|\lambda| \leqq \lambda_2$, $\lambda \neq 0$, $|\varrho| \leqq \varrho_2$

$$\operatorname{sgn}\,\{-\alpha(\lambda)/(\alpha_3(\lambda) + l(2\pi; \varrho, \lambda))\} = \operatorname{sgn}\,\{-\alpha(\lambda)/\alpha_3(0)\}\,. \tag{2.31}$$

Aus der Stetigkeit der Funktionen $\gamma_1(\lambda)$ und $\gamma_2(\lambda)$ und der Relation (2.21) folgt die Existenz einer positiven Zahl $\varrho_3$, $\varrho_3 \leqq \varrho_2$, so daß für $|\lambda| \leqq \lambda_2$ und $|\varrho| \leqq \varrho_3$ die Relation $\gamma_1(\lambda) + \gamma_2(\lambda)\varrho \geqq c_3 > 0$ gilt. Folglich haben wir für $|\lambda| \leqq \lambda_2$, $\lambda \neq 0$, $|\varrho| \leqq \varrho_3$ nach (2.31)

$$\operatorname{sgn}\,\{-\alpha(\lambda)\,[\gamma_1(\lambda) + \gamma_2(\lambda)\,\varrho]/(\alpha_3(\lambda) + l(2\pi; \varrho, \lambda))\} = \operatorname{sgn}\,\{-\alpha(\lambda)/\alpha_3(0)\}\,.$$

Die Relation $\alpha_3(0)\,\alpha(\lambda) < 0$ ist somit eine notwendige Bedingung für die Existenz reeller Nullstellen von $\tilde{\psi}(\varrho, \lambda)$. Aus der Definition von $g(\varrho, \lambda)$ folgt unmittelbar, daß eine positive Zahl $\lambda_3$ existiert, $\lambda_3 \leqq \lambda_2$, so daß $g(\varrho, \lambda)$ für $|\lambda| \leqq \lambda_3$ das Intervall $[0, \varrho_3]$ stetig in sich abbildet. Folglich besitzt die Gleichung $\varrho = g(\varrho, \lambda)$ für $|\lambda| \leqq \lambda_3$ unter der Voraussetzung $\alpha_3(0)\,\alpha(\lambda) < 0$ im Intervall $0 \leqq \varrho \leqq \varrho_3$ eine positive Nullstelle. Ein analoges Resultat gilt für die Gleichung $\varrho = -g(\varrho, \lambda)$ im Intervall $-\varrho_2 \leqq \varrho \leqq 0$. Da es höchstens zwei Nullstellen im Intervall $|\varrho| \leqq \varrho_2$ gibt, ist das Lemma bewiesen.

Aus den obigen Untersuchungen ergibt sich unmittelbar das folgende Resultat:

Satz 2.1: *Es seien die Voraussetzungen $V_1$ bis $V_4$ erfüllt. Falls die Funktion $\alpha(\lambda)$ an der Stelle $\lambda = 0$ ihr Vorzeichen ändert, besitzen alle Systeme (2.1) für $\lambda$ aus einer halbseitigen Umgebung $\Lambda^*$ von $\lambda = 0$ in einer hinreichend kleinen Umgebung von $x = 0$ genau einen Grenzzyklus $\Gamma(\lambda)$, dessen Stabilitätsverhalten zu dem des singulären Punktes $x = 0$ entgegengesetzt ist. Die Bifurkationsrichtung wird durch $\operatorname{sgn}\,[-\alpha(\lambda_1)\,\alpha_3(0)]$ bestimmt, d. h., bei positivem (negativem) Vorzeichen besteht $\Lambda^*$ aus positiven (negativen) $\lambda$-Werten. Ändert die Funktion $\alpha(\lambda)$ für $\lambda = 0$ ihr Vorzeichen nicht und gilt $\alpha(\lambda_1)\,\alpha_3(0) < 0$ ($> 0$), dann besitzen alle Systeme (2.1) für $\lambda$ aus einer hinreichend kleinen Umgebung von $\lambda = 0$ in einer hinreichend kleinen Umgebung von $x = 0$ genau einen (keinen) Grenzzyklus $\Gamma(\lambda)$. Falls $\Gamma(\lambda)$ existiert, ist sein Stabilitätsverhalten entgegengesetzt dem des singulären Punktes $x = 0$.*

3. Als Anwendungsbeispiel betrachten wir das Differentialgleichungssystem

$$\frac{dx_1}{dt} = x_2, \qquad \frac{dx_2}{dt} = -x_1 - \sigma(\lambda)\, x_2 + a x_1^2 + c x_1 x_2 + b x_2^2, \tag{3.1}$$

wobei $\sigma(\lambda)$ durch

$$\sigma(\lambda) = \begin{cases} d_1\lambda & \text{für } \lambda \geqq 0, \\ d_2\lambda & \text{für } \lambda \leqq 0 \end{cases} \tag{3.2}$$

definiert ist. Über die auftretenden Koeffizienten setzen wir die Gültigkeit der Relationen

$$d_1 d_2 > 0, \qquad c(a + b) \neq 0 \qquad (3.3),\ (3.4)$$

voraus. Der Spezialfall $d_1 = d_2 = 1$, d. h. $\sigma(\lambda) = \lambda$, ist in der Theorie der autonomen Schwingungen von Interesse und wurde erstmalig von N. N. BAUTIN [4] untersucht.

Satz 3.1: *Unter den Voraussetzungen* (3.3) *und* (3.4) *zweigt beim Durchgang von $\lambda$ durch $\lambda = 0$ vom singulären Punkt $x_1 = x_2 = 0$ genau ein Grenzzyklus $\Gamma(\lambda)$ ab. Die Bifurkationsrichtung wird durch die Relation* sgn $[-d_1 c(a + b)]$ *bestimmt. Das Stabilitätsverhalten von $\Gamma(\lambda)$ ist entgegengesetzt dem des singulären Punktes $x_1 = x_2 = 0$ des Systems* (3.1).

Beweis: Aus (3.1) folgt unmittelbar, daß die in $V_1$ geforderten Regularitätsvoraussetzungen für alle $x$ und $\lambda$ erfüllt sind. Die Gültigkeit der Relation (2.2) läßt sich ebenfalls unmittelbar verifizieren. Die zum Punkt $x = 0$ gehörenden Eigenwerte $\lambda_{1,2}$ der JACOBI-Matrix haben die Gestalt

$$\lambda_{1,2} = \alpha(\lambda) \pm i\beta(\lambda) = -\sigma(\lambda)/2 \pm i\sqrt{1 - \sigma^2(\lambda)/4}\,.$$

Setzen wir $d = \max\,(|d_1|, |d_2|)$, so sind die Relationen (2.3) und (2.4) für $|\lambda| < 2/d$ erfüllt. Ersetzen wir in (3.1) $t$ durch $-t$, so erhalten wir das System

$$\frac{dx_1}{dt} = -x_2, \qquad \frac{dx_2}{dt} = x_1 + \sigma(\lambda)\, x_2 - a x_1^2 - c x_1 x_2 - b x_2^2. \tag{3.5}$$

Es läßt sich leicht verifizieren, daß das System (3.5) die Bedingungen (2.25), (2.26) erfüllt, wobei die Relationen

$$\beta(0) = 1, \qquad \alpha_3(0)_{|(3.5)} = -\frac{\pi}{4}\, c(a + b)$$

gelten. Unter Berücksichtigung von Bemerkung 2.3.2 erhalten wir

$$\alpha_3(0)_{|(3.1)} = \frac{\pi}{4}\, c(a + b)\,.$$

Nach (3.4) stellt folglich $x_1 = x_2 = 0$ für $\lambda = 0$ einen strukturell instabilen Fokus der Vielfachheit 1 für das System (3.1) dar. Da $\alpha(\lambda) \equiv -\sigma(\lambda)/2$ an der Stelle $\lambda = 0$ das Vorzeichen ändert, erhalten wir Satz 3.1 durch eine unmittelbare Anwendung von Satz 2.1. q. e. d.

## Literatur


1 ANDRONOV, A. A., Gesammelte Werke (russ.). Moskau 1956.
2 ANDRONOV, A. A.; LEONTOVIČ, E. A.; GORDON, I. I.; MAIER, A. G., Bifurkationstheorie dynamischer Systeme in der Ebene (russ.), Moskau 1967.
3 BANKS, H. T., Modeling and Control in the Biomedical Sciences, Lecture Notes in Biomathematics Vol. 6, Berlin Springer, 1975.
4 BAUTIN, N. N., Über eine Differentialgleichung, die einen Grenzzyklus besitzt (russ.). J. techn. fiz. 9, 601—611 (1939).
5 HOPF, E., Abzweigung einer periodischen Lösung von einer stationären Lösung eines Differentialsystems, Ber. Sächs. Akad. Wiss., Math.-Phys. Klasse 94, 1—22 (1942).
6 HSÜ, I. D.; KAZARINOFF, N. D.; An applicable HOPF bifurcation formula and instability of small periodic solutions of the FIELD-NOYES model, J. Math. Analysis Appl. 55, 61—89 (1976).
7 KELLEY, A., The stable, center-stable, center-unstable, unstable manifolds, J. Diff. Equs. 3, 546—570 (1967).
8 KNOBLOCH, H. W., HOPF Bifurcation via integral manifolds, Abh. AdW der DDR N 3, 413—429 (1977).
9 MARSDEN, J. E.; MCCRACKEN, M.; The HOPF Bifurcation and Its Applications, Applied Math. Sci., Vol. 19, New York 1976.
10 POINCARÉ, H., Les méthodes nouvelles de la Mécanique céleste. Vol. 1, Paris 1892.
11 POORE, A. B., On the theory and application of the HOPF-FRIEDRICHS bifurcation theory, Arch. Rat. Mech. Analysis 60, 371—393 (1976).
12 SAVAGEAU, M. A., Biochemical System Analysis, Reading 1976.
13 SCHNEIDER, K. R., Nichtlineare Differentialgleichungssysteme als mathematische Modelle neurophysiologischer Vorgänge. Dissertation, Humboldt-Universität Berlin 1967.
14 SCHNEIDER, K. R., HOPF bifurcation and integral manifolds, Preprint P-39/79 des ZIMM der AdW.
15 SCHNEIDER, K. R.; WEGNER, B., Asymptotic behaviour of the period of a closed trajectory, Abh. AdW der DDR N 4, 215—219 (1977).
16 Some Mathematical Questions in Biology. VII. Lectures on Mathematics in the Life Sciences. Vol. 8, AMS, Providence 1976.
17 TYSON, JOHN J., The BELOUSOV-ZHABOTINSKII Reaction. Lecture Notes in Biomathematics Vol. 10, Springer, Berlin 1976.


Eingereicht: 24. 11. 1978, korrigierte Fassung 16. 7. 1979

*Anschrift:* Dr. habil. KLAUS R. SCHNEIDER, ZIMM der AdW der DDR, Rudower Chaussee 5, DDR-1199 Berlin

ZAMM 60, 25 – 30 (1980)

K. GRÖGER

# Asymptotic Behaviour of Elastoplastic Bodies under Periodic Loading

In dieser Arbeit beweisen wir, daß jeder quasi-statische Deformationsprozeß eines elastisch-plastischen Körpers mit isotroper und kinematischer Verfestigung gegen einen periodischen Prozeß konvergiert, wenn die Belastung zeitlich periodisch ist.

In this paper we prove that every quasi-static deformation process of an elastoplastic body with isotropic and kinematic hardening tends to a periodic process if the loading is periodic in time.

В данной работе мы доказываем, что любой квазистатический процесс деформирования упруго-пластического тела с изотропным и кинематическим упрочнением сходится к некоторому периодическому процессу, если нагрузка периодична во времени.

## Introduction

In this paper we deal with quasi-static deformation processes of elastoplastic bodies subject to periodic loading. We assume that the hardening rule of the bodies describes a combination of isotropic and kinematic hardening. The paper consists of three sections. In the first section we introduce notations and we formulate the equations characterizing quasi-static processes of elastoplastic bodies. In the second section we prove that for every periodic loading process there exist periodic deformation processes. We show that all periodic processes are necessarily purely elastic due to the isotropic hardening effect. In the third section of the paper we prove that all deformation processes subject to a periodic loading tend in a certain sense to a periodic process as $t \to \infty$. This is the main result of the paper. Our results are somewhat better than those obtained by HALPHEN [4] in the case of purely kinematic hardening.

We have restricted our considerations to three-dimensional bodies in order to avoid unnecessary complications in the presentation of the results. In the same way it is possible to treat other problems such as the torsion of elastoplastic cylindrical bars.

## 1. Formulation of the problem

We shall use the following notations:

$G \subset \mathbb{R}^3$ bounded domain occupied by the body under consideration;
$\Gamma$ part of the boundary $\partial G$ where the body is clamped (the measure of which is assumed to be positive);
$U := \{u \in H^1(G; \mathbb{R}^3) \mid u_\Gamma = 0\}$ space of virtual displacements;
$U'$ space of forces (dual to $U$);
$\mathscr{S}$ HILBERT space of symmetric $3 \times 3$ matrices;
$Y := L^2(G; \mathscr{S})$ HILBERT space of square integrable functions on $G$ with values in $\mathscr{S}$, identified with its dual (space of stress and strain tensor fields);
$Z := L^2(G)$ space of internal parameters, identified with its dual;
$A \in \mathscr{L}(Y; Y)$ continuous linear operator characterizing HOOKE's law;
$B \in \mathscr{L}(Y; Y)$ continuous linear operator characterizing the kinematic hardening of the body;
$C \subset Y \times Z$ convex set characterizing the hardening rule of the body.

We shall assume that $A$, $B$ and $C$ satisfy the following conditions:

$$A, B \in \mathscr{L}(Y; Y) \quad \text{are symmetric and positive definite;} \tag{1}$$

$$C = \{(\tau, \alpha) \in Y \times Z \mid F(\tau(x)) \leqq \alpha(x) \quad \text{for a.e. } x \in G\}\,, \tag{2}$$

where

$$F: \mathscr{S} \to \mathbb{R} \text{ is convex and continuously differentiable such that } F' \text{ is bounded on } \mathscr{S}\,. \tag{3}$$

By $Ku$ we denote the infinitesimal strain field corresponding to a displacement $u \in U$, i.e., we define $K \in \mathscr{L}(U; Y)$ by

$$\forall u \in U: \quad (Ku)_{ij} := \tfrac{1}{2}\,(u_{i,j} + u_{j,i})\,. \tag{4}$$

The operator $K$ satisfies KORN's inequality, i.e.

$$\forall u \in U: \|Ku\|_Y \geqq k_0\,\|u\|_U\,, \qquad k_0 > 0$$

(see HLAVÁČEK-NEČAS [6]).

If $S$ is a compact interval of time and $X$ any HILBERT space we denote by $L^2(S; X)$ the HILBERT space of square integrable mappings from $S$ into $X$ with its usual scalar product. If $u \in L^2(S; X)$ we denote by $u'$ the derivative of $u$ with respect to time in the sense of $X$-valued distributions. We define

$$H^1(S; X) := \{u \in L^2(S; X) \mid u' \in L^2(S; X)\}\,.$$

This space equipped with the scalar product

$$(u, v)_{H^1(S;\, X)} := (u, v)_{L^2(S;\, X)} + (u', v')_{L^2(S;\, X)}$$

is a HILBERT space. The space $L^2(S; L^2(G))$ will be identified in the usual way with $L^2(S \times G)$. Therefore, if $u \in L^2(S; L^2(G))$ instead of $(u(t))\,(x)$ we use the simpler notation $u(t, x)$.

Let now $S := [0, T]$, $T > 0$, be a fixed compact interval of time and let $f\colon \mathbb{R}^+ \to U'$ be a function such that

$$\forall t \in \mathbb{R}^+: \quad f(t + T) = f(t)\,, \qquad f \in H^1([0, t]; U')\,. \tag{5}$$

This function represents a loading of the body periodic in time. In what follows we assume the conditions (1)—(5) to be satisfied without mentioning this explicitely.

A quasi-static deformation process of the elastoplastic body under consideration caused by the periodic loading $f$ can be described by functions $u, \sigma, p, \alpha$ (displacement, stress, plastic strain, internal parameter) satisfying the following conditions (see e.g. GRÖGER [3]):

$$\forall t \in \mathbb{R}^+: \quad Ku(t) = p(t) + A^{-1}\sigma(t)\,, \qquad K^*\sigma(t) = f(t)\,; \tag{6}$$

$$(p'(t), -\alpha'(t)) \in \partial I_C(\sigma(t) - Bp(t), \alpha(t)) \quad \text{for a.e. } t \in \mathbb{R}^+; \tag{7}$$

$$(u, \sigma, p, \alpha) \in H^1([0, t]; U \times Y \times Y \times Z) \quad \text{for all } t > 0\,, \tag{8}$$

where $K^* \in \mathcal{L}(Y; U')$ denotes the adjoint operator of $K$ and $\partial I_C$ is the subdifferential mapping of the indicator functional $I_C$ of the convex set $C$ (for the definition of $\partial I_C$ see BRÉZIS [1]). The relations (6) are the compatibility condition and the equilibrium condition whereas (7) is a constitutive relation of the material. This type of relation was introduced by NGUYEN [7] and HALPHEN-NGUYEN [5]. The relation (7) can be written in the following equivalent way:

$$\begin{aligned} &\forall (\bar\tau, \bar\alpha) \in C: \quad \big(p'(t),\ \bar\tau - (\sigma(t) - Bp(t))\big)_Y - \big(\alpha'(t), \bar\alpha - \alpha(t)\big)_Z \leqq 0 \\ &\text{for a.e. } t \in \mathbb{R}^+; \qquad \forall t \in \mathbb{R}^+: \ \big(\sigma(t) - Bp(t)\,,\ \alpha(t)\big) \in C\,. \end{aligned} \tag{9}$$

From results obtained in [3] it follows immediately

Theorem 1. *If* $(u_0, \sigma_0, p_0, \alpha_0) \in U \times Y \times Y \times Z$ *are initial values such that*

$$Ku_0 = p_0 + A^{-1}\sigma_0\,, \qquad K^*\sigma_0 = f(0)\,, \qquad (\sigma_0 - Bp_0, \alpha_0) \in C$$

*then there exists a unique solution* $(u, \sigma, p, \alpha)$ *to problem* (6)—(8) *satisfying the condition*

$$(u, \sigma, p, \alpha)(0) = (u_0, \sigma_0, p_0, \alpha_0)\,.$$

In the following sections we shall deal with periodic solutions to problem (6)—(8) and with the asymptotic behaviour of arbitrary solutions to this problem. To this end another equivalent formulation of problem (6)—(8) turns out to be quite useful. Let the operators $P \in \mathcal{L}(Y; Y)$, $Q \in \mathcal{L}(Y; Y)$ and $R \in \mathcal{L}(U'; Y)$ be defined by

$$R := AK(K^*AK)^{-1}\,, \qquad Q := RK^*\,, \qquad P := I - Q \qquad (I \text{ identity map of } Y)\,. \tag{10}$$

The operator $R$ is defined in such a way that $Rf$ is the stress which would correspond to the force $f$ if the body were purely elastic with HOOKE's tensor $A$. It is easy to check that $P^2 = P$, $Q^2 = Q$ and $\operatorname{Im} P = \operatorname{Ker} K^*$, $\operatorname{Im} Q = \operatorname{Im} AK$, where Im and Ker denote the image and the kernel of a linear operator. Moreover, it is easy to see that $PA \in \mathcal{L}(Y; Y)$ is symmetric and positive. Hence $L := PA + B \in \mathcal{L}(Y; Y)$ is symmetric and positive definite. From (6) it follows

$$\forall t \in \mathbb{R}^+: \quad Q\sigma(t) = Rf(t)\,, \qquad P\sigma(t) = -PAp(t)\,.$$

Thus

$$\forall t \in \mathbb{R}^+: \quad \sigma(t) = Rf(t) - PAp(t)\,, \qquad \sigma(t) - Bp(t) = Rf(t) - Lp(t)\,. \tag{11}$$

Consequently, from (6)—(8) it follows

$$(p'(t), -\alpha'(t)) \in \partial I_C(Rf(t) - Lp(t), \alpha(t)) \quad \text{for a.e. } t \in \mathbb{R}^+ \tag{12}$$

$$(p, \alpha) \in H^1([0, t]; Y \times Z) \quad \text{for all } t > 0\,.$$

Conversely, let $(p, \alpha)$ be a solution to (12). It is easy to see that to the given functions $p$ and $\alpha$ there exists a unique solution $(u, \sigma, p, \alpha)$ to problem (6)—(8). This solution can be determined by (cf. (11) and (6)):

$$\forall t \in \mathbb{R}^+: \quad \sigma(t) := Rf(t) - PAp(t)\,, \qquad u(t) := (K^*AK)^{-1}\,(f(t) + K^*Ap(t))\,. \tag{13}$$

Therefore problem (12) may be regarded as an equivalent formulation of problem (6)—(8).

## 2. Periodic Solutions

Theorem 2. *For every periodic solution* $(u, \sigma, p, \alpha)$ *to problem* (6)—(8) *there exist elements* $p_0 \in Y$ *and* $\alpha_0 \in Z$ *such that*

$$\begin{aligned} &\forall t \in \mathbb{R}^+: \quad u(t) = (K^*AK)^{-1}\,(f(t) + K^*Ap_0)\,, \\ &\sigma(t) = Rf(t) - PAp_0\,, \qquad p(t) = p_0\,, \qquad \alpha(t) = \alpha_0\,. \end{aligned} \tag{14}$$

The proof of Theorem 2 is based on the following lemma.

Lemma 1. *Let $(u, \sigma, p, \alpha)$ be a solution to problem (6)—(8). Then for all $t_0, t_1 \in \mathbb{R}^+$ such that $t_0 \leqq t_1$ we have $\alpha(t_0) \leqq \alpha(t_1)$ with respect to the usual ordering of $Z = L^2(G)$.*

Proof. Let $\beta \in Z$ be such that $\beta \geqq 0$. Then

$$\forall t \in \mathbb{R}^+: \quad (\sigma(t) - Bp(t), \alpha(t) + \beta) \in C .$$

Hence from (9) it follows

$$(\alpha'(t), \beta)_Z \geqq 0 \quad \text{for a.e. } t \in \mathbb{R}^+ .$$

By integration we find

$$(\alpha(t_1) - \alpha(t_0), \beta)_Z \geqq 0 .$$

Because $\beta \geqq 0$ is arbitrary this proves the assertion of the lemma.

Proof of Theorem 2. Lemma 1 shows that if $(u, \sigma, p, \alpha)$ is a periodic solution to problem (6)—(8) then $\alpha$ is constant in time. Let $\alpha(t) = \alpha_0$ for all $t \in \mathbb{R}^+$. For every $\tau \in Y$ the pair $(\tau, F(\tau(\cdot)))$ belongs to $C$. Therefore from (9) it follows

$$\forall \tau \in Y: \quad (p'(t), \tau - (\sigma(t) - Bp(t)))_Y \leqq 0 \quad \text{for a.e. } t \in \mathbb{R}^+ .$$

This is possible only if $p$ is also constant in time. Let $p(t) = p_0$ for all $t \in \mathbb{R}^+$. The remaining assertions of Theorem 2 follow from (13).

Theorem 3. *For every $p_0 \in Y$ there exists a periodic solution to problem (6)—(8) such that (14) holds.*

The proof of this theorem is based on the following simple lemma.

Lemma 2. *For every $g \in H^1(S; L^2(G))$ there exists a function $\alpha \in L^2(G)$ such that*

$$\forall t \in S: \quad |g(t, x)| \leqq \alpha(x) \quad \text{for a.e. } x \in G .$$

Proof. For every $t \in S$ we have

$$g(t) = g(0) + \int_0^t g'(s)\, ds .$$

Using the identification of $L^2(S; L^2(G))$ and $L^2(S \times G)$ we can write this as follows:

$$g(t, x) = g(0, x) + \int_0^t g'(s, x)\, ds \quad \text{for a.e. } x \in G .$$

Defining

$$\alpha(x) := |g(0, x)| + \int_S |g'(s, x)|\, ds \quad \text{for a.e. } x \in G$$

we obtain

$$|g(t, x)| \leqq |g(0, x)| + \int_0^t |g'(s, x)|\, ds \leqq \alpha(x) \quad \text{for a.e. } x \in G .$$

Since

$$\int_G |\alpha(x)|^2\, dx \leqq 2 \int_G \left\{|g(0, x)|^2 + \left(\int_S |g'(s, x)|\, ds\right)^2\right\} dx \leqq \text{const } \|g\|_{H^1(S;L^2(G))}$$

this proves the lemma.

Proof of Theorem 3. Let $p_0 \in Y$ be given arbitrarily. Because of the properties of the function $F$ there exist $M > 0$ and $\alpha_1 \in Z$ such that

$$\forall t \in S: \quad F((Rf(t) - Lp_0)(x)) \leqq M\, |(Rf(t))(x)| + \alpha_1(x) \quad \text{for a.e. } x \in G .$$

Applying Lemma 2 to the components of $Rf$ we see that there exists $\alpha_0 \in Z$ such that

$$\forall t \in S: \quad F((Rf(t) - Lp_0)(x)) \leqq \alpha_0(x) \quad \text{for a.e. } x \in G ,$$

i.e.

$$\forall t \in S: \quad (Rf(t) - Lp_0, \alpha_0) \in C .$$

Consequently,

$$\forall t \in \mathbb{R}^+: \quad (0, 0) \in \partial I_C(Rf(t) - Lp_0, \alpha_0) ,$$

i.e., the relation (12) holds if we define

$$\forall t \in \mathbb{R}^+: \quad (p, \alpha)(t) := (p_0, \alpha_0) .$$

As mentioned already in Section 1 the relation (12) is equivalent to problem (6)—(8). Thus we proved the existence of a solution to this problem corresponding to the given $p_0 \in Y$. From (13) it follows that this solution is periodic.

### 3. Asymptotic Behaviour

Throughout this section we assume that we are given a solution $(u, \sigma, p, \alpha)$ to problem (6)—(8) and we define

$$(u_n, \sigma_n, p_n, \alpha_n) \in H^1(S; U \times Y \times Y \times Z), \qquad n = 0, 1, \ldots,$$

by

$$\forall t \in S: \quad (u_n, \sigma_n, p_n, \alpha_n)(t) := (u, \sigma, p, \alpha)(t + nT).$$

Theorem 4. *There exists a periodic solution $(\tilde{u}, \tilde{\sigma}, \tilde{p}, \tilde{\alpha})$ to problem* (6)—(8) *such that*

$$p(t) \rightharpoonup \tilde{p}_0 \text{ in } Y, \qquad \alpha(t) \to \tilde{\alpha}_0 \text{ in } Z \text{ as } t \to \infty;$$

$$\left.\begin{array}{l} (u_n, \sigma_n, p_n) \rightharpoonup (\tilde{u}, \tilde{\sigma}, \tilde{p}) \text{ in } H^1(S; U \times Y \times Y) \\ (u_n', \sigma_n', p_n') \to (\tilde{u}', \tilde{\sigma}', 0) \text{ in } L^2(S; U \times Y \times Y) \\ \alpha_n \rightharpoonup \tilde{\alpha} \text{ in } H^1(S; Z) \end{array}\right\} \text{as } n \to \infty;$$

*where $\tilde{p}_0$ and $\tilde{\alpha}_0$ denote the value of $\tilde{p}$ and $\tilde{\alpha}$, respectively.*

The proof of this theorem will be divided into several lemmas. We shall use the following notation:

$$\forall p, \forall q \in Y: \quad (p, q)_L := (p, Lq)_Y, \qquad \|p\|_L := (p, p)_L^{1/2}.$$

Since $L \in \mathcal{L}(Y; Y)$ is symmetric and positive definite $\|\cdot\|_L$ is a norm on $Y$ which is equivalent to the usual norm on this space.

Lemma 3. *Let $(\tilde{u}, \tilde{\sigma}, \tilde{p}, \tilde{\alpha})$ be any periodic solution to problem* (6)—(8) *and let $(\tilde{p}, \tilde{\alpha})(t) = (p_0, \alpha_0)$ for $t \in \mathbb{R}^+$. Then, if $t \geqq t_0 \geqq 0$ we have*

$$\|p(t) - \tilde{p}_0\|_L^2 + \|\alpha(t) - \tilde{\alpha}_0\|_Z^2 \leqq \|p(t_0) - \tilde{p}_0\|_L^2 + \|\alpha(t_0) - \tilde{\alpha}_0\|_Z^2.$$

Proof. Applying (12) to $(p, \alpha)$ and to $(\tilde{p}, \tilde{\alpha})$ and using the monotonicity of $\partial I_C$ we obtain

$$0 \leqq \int_{t_0}^{t} \{(p'(s), - L(p(s) - \tilde{p}_0))_Y - (\alpha'(s), \alpha(s) - \tilde{\alpha}_0)_Z\} \, ds$$

$$= \tfrac{1}{2}\{\|p(t_0) - \tilde{p}_0\|_L^2 + \|\alpha(t_0) - \tilde{\alpha}_0\|_Z^2 - \|p(t) - \tilde{p}_0\|_L^2 - \|\alpha(t) - \tilde{\alpha}_0\|_Z^2\}.$$

This proves the lemma.

Lemma 4. *We have*

$$\|p'(t)\|_L^2 + \|\alpha'(t)\|_Z^2 = (p'(t), Rf'(t))_Y \quad \textit{for a.e. } t \in \mathbb{R}^+.$$

Proof. In view of Lemma 3.3 in BRÉZIS [1] the assertion follows immediately from (12).

Lemma 5. *There exists a function $\tilde{\alpha}$ which is constant in time such that*

$$\alpha_n \rightharpoonup \tilde{\alpha} \quad \textit{in } H^1(S; Z) \textit{ as } n \to \infty.$$

Proof. 1. Lemma 1 shows that there exists a subset $N$ of $G$ of measure 0 such that

$$\forall x \in G \setminus N: \quad \alpha_m(0, x) \leqq \alpha_n(0, x) \quad \text{if} \quad m, n \in \mathbb{N}, \quad m \leqq n.$$

Thus for $x \in G \setminus N$ there exists the limit

$$\tilde{\alpha}_0(x) := \lim_{n \to \infty} \alpha_n(0, x), \tag{15}$$

where it is not excluded that this limit is $+\infty$. For fixed $t \in S$ we have

$$\forall n \in \mathbb{N}: \quad \alpha_n(0, x) \leqq \alpha_n(t, x) \leqq \alpha_{n+1}(0, x) \quad \text{for a.e. } x \in G. \tag{16}$$

Consequently

$$\forall t \in S: \quad \tilde{\alpha}_0(x) = \lim_{n \to \infty} \alpha_n(t, x) \quad \text{for a.e. } x \in G. \tag{17}$$

2. From Lemma 3 and Lemma 4 it follows that the sequence $(\alpha_n)$ is bounded in $H^1(S; Z)$. If $(\alpha_{n_j})$ is a subsequence converging weakly in $H^1(S; Z)$ then for every $t \in S$ the sequence $(\alpha_{n_j}(t))$ converges weakly in $Z$. On account of (17) the limit of this sequence is necessarily $\tilde{\alpha}_0$ (cf. [2], Lemma 1.19, Ch. I). Hence $\tilde{\alpha}_0 \in Z$ and $\alpha_n \rightharpoonup \tilde{\alpha}$ in $H^1(S; Z)$ if we denote by $\tilde{\alpha}$ the constant function with the value $\tilde{\alpha}_0$.

Lemma 6. *We have*

$$\alpha(t) \to \tilde{\alpha}_0 \quad \textit{in } Z \textit{ as } t \to \infty,$$

*where $\tilde{\alpha}_0$ denotes the function defined by* (15).

Proof. Let $(t_k)$ be a sequence of positive numbers converging to $+\infty$. In the same way as (17) one can prove that

$$\lim_{k \to \infty} \alpha(t_k, x) = \tilde{\alpha}_0(x) \quad \text{for a. e. } x \in G.$$

Furthermore

$$\alpha(0, x) \leqq \alpha(t_k, x) \leqq \tilde{\alpha}_0(x) \quad \text{for a.e. } x \in G .$$

Thus LEBESGUE's Theorem yields

$$\alpha(t_k) \to \tilde{\alpha}_0 \quad \text{in } Z \text{ as } k \to \infty .$$

Since $(t_k)$ is an arbitrary sequence converging to $+\infty$ this proves the lemma.

Lemma 7. *We have*

$$\alpha_n \to \tilde{\alpha} \quad in \quad L^2(S; Z) \quad as\ n \to \infty ,$$

*where $\tilde{\alpha}$ is the function introduced in Lemma 5.*

Proof. In view of (16) and Lemma 6 we have

$$\lim_{n\to\infty} \int_S \int_G |\alpha_n(t, x) - \tilde{\alpha}_0(x)|^2 \,dx\, dt \leqq \lim_{n\to\infty} \int_S \int_G |\alpha_n(0, x) - \tilde{\alpha}_0(x)|^2 \,dx\, dt = \lim_{n\to\infty} T\, \|\alpha(nT) - \tilde{\alpha}_0\|_Z^2 = 0 .$$

Lemma 8. *We have*

$$p_n' \rightharpoonup 0 \quad in \quad L^2(S; Y) \quad as\ n \to \infty .$$

Proof. Let $\bar{\tau} \in L^2(S; Y)$ be arbitrary and let

$$\forall t \in S\colon \bar{\alpha}(t) := F(\bar{\tau}(t, \cdot)) .$$

Then

$$\forall t \in S\colon \quad (\bar{\tau}(t), \bar{\alpha}(t)) \in C .$$

Hence from (9) and (11) it follows

$$\int_S \{(p_n'(t), \bar{\tau}(t) - Rf(t) + Lp_n(t))_Y - (\alpha_n'(t), \bar{\alpha}(t) - \alpha_n(t))_Z\}\, dt \leqq 0 .$$

Because of the properties of the sequence $(\alpha_n)$ proved in Lemma 5 and Lemma 7 this implies that

$$\overline{\lim_{n\to\infty}} \int_S (p_n'(t), \bar{\tau}(t))_Y\, dt \leqq \overline{\lim_{n\to\infty}} \int_S (p_n'(t), Rf(t) - Lp_n(t))_Y\, dt .$$

The right hand side of this inequality is finite (cf. Lemma 3, Lemma 4) and independent of $\bar{\tau}$. Since $\bar{\tau} \in L^2(S; Y)$ was arbitrary this is possible only if the sequence $(p_n')$ converges weakly to 0 in $L^2(S; Y)$.

Lemma 9. *We have*

$$(u_n', \sigma_n', p_n', \alpha_n') \to ((K^*AK)^{-1} f', Rf', 0, 0) \quad in \quad L^2(S; U \times Y \times Y \times Z) .$$

Proof. From Lemma 4 it follows

$$\int_S \{\|p_n'(t)\|_L^2 + \|\alpha_n'(t)\|_Z^2\}\, dt = \int_S (p_n'(t), Rf'(t))_Y\, dt .$$

In view of Lemma 8 the right hand side converges to 0 as $n \to \infty$. Thus

$$(p_n', \alpha_n') \to (0, 0) \quad \text{in} \quad L^2(S; Y \times Z) .$$

The remaining assertions of the lemma follow from this result and (13).

Lemma 10. *Let $(p_{n_j}(0))$, $j = 1, 2, \ldots$, be a subsequence of $(p_n(0))$ such that*

$$p_{nj} \rightharpoonup \tilde{p}_0 \quad in \quad Y\ as\ j \to \infty .$$

*If we define*

$$\forall t \in \mathbb{R}^+\colon \tilde{u}(t) := (K^*AK)^{-1} (f(t) - K^*A\tilde{p}_0), \qquad \tilde{\sigma}(t) := Rf(t) - PA\tilde{p}_0, \qquad p(t) := \tilde{p}_0 ,$$

*then $(\tilde{u}, \tilde{\sigma}, \tilde{p}, \tilde{\alpha})$ is a periodic solution to problem (6)—(8) and we have*

$$(u_{n_j}, \sigma_{n_j}, p_{n_j}) \rightharpoonup (\tilde{u}, \tilde{\sigma}, \tilde{p}) \quad in \quad H^1(S; U \times Y \times Y) \quad as\ j \to \infty . \tag{18}$$

Proof. 1. Let $\tau \in L^2(S; Y)$ be arbitrary. In view of $p_n' \rightharpoonup 0$ in $L^2(S; Y)$ it holds

$$\lim_{j\to\infty} \int_S (p_{n_j}(t), \tau(t))_Y\, dt = \lim_{j\to\infty} \int_S (p_{n_j}'(t), \int_t^T \tau(s)\, ds)_Y\, dt + (p_{n_j}(0), \int_S \tau(s)\, ds)_Y$$

$$= (\tilde{p}_0, \int_S \tau(s)\, ds)_Y = \int_S (\tilde{p}(s), \tau(s))_Y\, ds .$$

This result along with the convergence of the sequence $(p_n')$ proves that $p_{n_j} \rightharpoonup \tilde{p}$ in $H^1(S; Y)$. The assertions concerning the convergence of the sequences $(u_{n_j})$ and $(\sigma_{n_j})$ follow from this relation in view of (13).

2. The first step of the proof shows that

$$\forall t \in S\colon \quad (Rf(t) - Lp_{n_j}(t), \alpha_{n_j}(t)) \rightharpoonup (Rf(t) - L\tilde{p}_0, \tilde{\alpha}_0) \quad \text{in} \quad Y \times Z .$$

Because of $(Rf(t) - Lp_{n_j}(t), \alpha_{n_j}(t)) \in C$ (cf. (12)) this implies that

$$\forall t \in S\colon \quad (Rf(t) - L\tilde{p}_0, \tilde{\alpha}_0) \in C . \tag{19}$$

This proves that $(\tilde{u}, \tilde{\sigma}, \tilde{p}, \tilde{\alpha})$ is a (periodic) solution to problem (6)—(8) (cf. the formulation (12) of this problem and the relations (13)).

**Remark.** To conclude the proof of Theorem 4 we have to show only that

$$p(t) \rightharpoonup \tilde{p}_0 \quad \text{in} \quad Y \quad \text{as} \quad t \to \infty. \tag{20}$$

Indeed, if (20) is valid then $p_n(0) \rightharpoonup \tilde{p}_0$ in $Y$ as $n \to \infty$ and we can use the results of Lemma 10 not only for a subsequence $(n_j)$ of natural numbers but for the whole sequence of natural numbers. The proof of (20) given below is an adaption of a proof given by PAZY [8] in connection with the asymptotic behaviour of semigroups of contractive operators.

**Lemma 11.** *There exists $\tilde{p}_0 \in Y$ such that*

$$p(t) \rightharpoonup \tilde{p}_0 \quad \textit{in} \quad Y \quad \textit{as} \quad t \to \infty.$$

**Proof.** Let

$$E := \{p_0 \in Y \mid \forall t \in S: (Rf(t) - Lp_0, \tilde{\alpha}_0) \in C\}.$$

Obviously, the set $E$ is convex and closed and it is not empty (cf. (19)). The assertion of Lemma 3 holds with $p_0 \in E$ instead of $\tilde{p}_0$. Therefore the function

$$t \to \|p(t) - p_0\|_L^2 + \|\alpha(t) - \tilde{\alpha}_0\|_Z^2$$

is not increasing in time. Since $\alpha(t) \to \tilde{\alpha}_0$ in $Z$ as $t \to \infty$ this implies that for every $p_0 \in E$ there exists the (finite) limit

$$\varrho(p_0) := \lim_{t \to \infty} \|p(t) - p_0\|_L.$$

Furthermore, let

$$\forall \varrho \in \mathbb{R}^+: E_\varrho := \{p_0 \in E \mid \varrho(p_0) \leqq \varrho\}.$$

It is easy to check that $E_\varrho$ is convex and closed and that it is nonempty for sufficiently large $\varrho$. Let $\varrho_0$ be the infimum of all $\varrho \in \mathbb{R}^+$ such that $E_\varrho \neq \emptyset$. The set $E_{\varrho_0}$ is not empty since $E_{\varrho_0} = \bigcap_{\varrho > \varrho_0} E_\varrho$ and the sets $E_\varrho$ are weakly compact and have the finite intersection property. Let $\tilde{p}_0 \in E_{\varrho_0}$. Let us assume that $p(t)$ does not converge weakly to $\tilde{p}_0$ in $Y$ as $t \to \infty$. Then there exists a sequence $(t_j)$ in $\mathbb{R}^+$ converging to $+\infty$ such that $p(t_j) \rightharpoonup p_0 \neq \tilde{p}_0$. If $t_j \in [n_j T, (n_j + 1) \times T[$ then $p_{n_j}(0) = p(n_j T) \rightharpoonup p_0$ because

$$p(t_j) - p_{n_j}(0) = \int_0^{t_j - n_j T} p'_{n_j}(s)\, ds \to 0 \quad \text{in} \quad Y$$

(cf. Lemma 9). Therefore it holds (19) with $p_0$ instead of $\tilde{p}_0$, i.e. $p_0 \in E$. Now

$$\|p(t_j) - \tilde{p}_0\|_L^2 = \|p(t_j) - p_0\|_L^2 + 2(p(t_j) - p_0, p_0 - \tilde{p}_0)_L + \|p_0 - \tilde{p}_0\|_L^2.$$

Letting $j \to \infty$ we find

$$\varrho_0^2 = (\varrho(p_0))^2 + \|p_0 - \tilde{p}_0\|_L^2.$$

This contradicts the definition of $\varrho_0$. Therefore $p(t) \rightharpoonup \tilde{p}_0$ in $Y$ as $t \to \infty$ and the proof is complete. (Simultaneously we have shown that $E_{\varrho_0}$ consists of a single element.)

**Remark.** If there exists a subsequence of $(p_n(0))$ converging strongly in $Y$ then

$$(u_n, \sigma_n, p_n) \to (\tilde{u}, \tilde{\sigma}, \tilde{p}) \quad \text{in} \quad H^1(S; U \times Y \times Y).$$

Indeed, in this case we have $\varrho(\tilde{p}_0) = 0$, i.e. $p(t) \to \tilde{p}_0$ in $Y$. In view of $p'_n \to 0$ in $L^2(S; Y)$ this implies that $p_n \to \tilde{p}$ in $H^1(S; Y)$. The remaining assertions follow from this relation and (13).

## References

1 H. BRÉZIS, Operateurs maximaux monotones et semi-groupes de contractions dans les espaces de HILBERT, North-Holland Mathematics Studies 5 (1973).

2 H. GAJEWSKI, K. GRÖGER, K. ZACHARIAS, Nichtlineare Operatorgleichungen und Operatordifferentialgleichungen, Berlin 1974.

3 K. GRÖGER, Evolution equations in the theory of plasticity, Proceedings of the Fifth Summer School on Nonlinear Operators, held at Berlin September 1977.

4 B. HALPHEN, L'accomodation des structures élastoplastiques à écrouissage cinématique, C.R. Acad. Sci. Paris Ser. A 283, 799—802 (1976).

5 B. HALPHEN, NGUYEN, Q. S., Sur les matériaux standards généralizés, J. Mécanique 14, 39—63 (1975).

6 I. HLAVÁČEK, J. NEČAS, On inequalities of KORN's type (I, II), Arch. for Rat. Mech. and Anal. 36, 305—311, 312—334 (1970).

7 NGUYEN, Q. S., Materiaux élastoplastiques écrouissables. Distribution de la contrainte dans une evolution quasi-statique, Arch. Mech. Stos. 25, 695—702 (1973).

8 A. PAZY, On the asymptotic behavior of semigroups of non-linear contractions in HILBERT space, J. Funct. Anal. 27, 292—307 (1978).

Eingereicht am 11. 12. 1978

*Anschrift*: Dr. KONRAD GRÖGER, Zentralinstitut für Mathematik und Mechanik der AdW der DDR, 108 Berlin, Mohrenstraße 39, DDR

ZAMM 60, 31 – 37 (1980)

L. M. DE SOCIO / L. MISICI

# Free Convection in a Trench with Radiative Wall Conditions

*Free convection is considered in a rectangular trench with an assigned temperature distribution at the bottom and radiative vertical walls. Following a domain perturbation technique, the problem reduces to the Stokes' equation. After proving the existence and uniqueness of the solution, from a functional point of view, the stream function is expressed in closed form via a bi-harmonic series and the shape of the free surface is also given. An application follows in order to show the convergence characteristics of the solution and the flow structure.*

*Die freie Konvektion wird in einer rechteckigen Rinne bei Temperaturverteilung am Boden und strahlenden vertikalen Wänden betrachtet. Unter Asunutzung eines Störungsansatzes reduziert sich das Problem auf die Stokes'-Gleichung. Nach dem funktionalanalytischen Beweis der Existenz und Einzigkeit der Lösung wird die Stromfunktion in geschlossener Form durch eine biharmonische Reihe ausgedrückt. Die Form der freien Oberfläche ist auch gegeben. Um die Konvergenzeigenschaften der Lösung und die Struktur der Strömung zu demonstrieren, folgt ein Anwendungsbeispiel.*

Рассматривается свободная конвекция в некотором желобке с прямоугольным сечением в условиях распределения температуры на дне и излучающих вертикальных стенок. С использованием некоторого метода возмущения задача сводится к решению уравнения Стокса. После доказательства существования и единственности решения функционально-аналитическим способом функция тока выражается в замкнутой форме с помощью бигармонического ряда. Форма свободной поверхности также задаётся. С целью иллюстрации свойств сходимости решения и структуры течения приводится один пример.

## 1. Introduction and Formulation of the Problem

In recent years an increasing effort has been devoted to the solution of the problem or recirculating natural convection in cavities. Furthermore, the situations in the presence of free surfaces received particular attention. These problems present noticeable interest from the point of view of the technical applications and for the noteworthy mathematical aspects. In general, only approximate solutions are available in analytic form [1, 2], while a large number of investigations is based on the numerical solution of the pertinent equations.

A complete review of the related literature would be extremely long. One should refer to some fundamental review works as those listed as [3] and [4] and to recent additions as [5, 6, 7].

In this paper, it will be shown how an analytic solution can be found to the fluid-dynamic aspects of the free convection in an open rectangular cavity. The investigated physical situation corresponds to a trench of infinite length, whose vertical axes are at a distance $2d$. In the channel a homogeneous, viscous liquid of density $\varrho$ and viscosity coefficient $\mu$ is contained.

The thermal diffusivity of the fluid is $k$, the coefficient of thermal expansion is $\beta$, while $\sigma$ is the surface tension coefficient. In the following, $g$ is the gravity constant, $T$ is the absolute temperature and $p$ the pressure.

Let $x$ and $y$ be a pair of Cartesian axes, with $x$ horizontal, perpendicular to the walls and corresponding to the free surface of the fluid at rest in isothermal conditions. The $y$ axis is directed along the vertical centerline of the trench and points positively upward. The open channel is of finite depth and the bottom is located at $y = -D$. At the vertical walls, "radiative" conditions exist, as expressed by $\pm\lambda\,[\partial T/\partial x] = h_T[T(\pm d, y) - T_\infty]$, at $x = \pm d$, where $T_\infty$ is the temperature of the surroundings, $h_T$ is the outside heat transfer coefficient and $\lambda$ is the thermal conductivity coefficient. In the above definition, the adjective "radiative" is conventionally used in classical textbooks [8] instead of the more significant word "convective" [9]. At the bottom, a continuous temperature distribution $T(x, -D)$ is prescribed, whereas at the free surface, whose equation $y = \varphi(x)$ is still an unknown of the problem, adiabatic conditions are assumed, i.e. $[\partial T/\partial y - \varphi'\,\partial T/\partial x] = 0$. This last circumstance corresponds to a negligible heat transfer coefficient between the recirculating liquid and the external gas at the interface.

The physical situation which has been just described belongs to the class of single phase open thermosyphon systems from the engineering point of view [4].

Some further nomenclature is necessary. Under the effects of the temperature gradients a natural convection field is established in the trench with the corresponding velocity $\boldsymbol{u}(x, y)$ and pressure $p(x, y)$ distributions. Let $u$ and $v$ be the $x$- and $y$-components of $\boldsymbol{u}$ and let one consider, for convenience, a reduced pressure $\Phi(x, y) = p(x, y) - p_a - \varrho g y$, where $p_a$ stays for the external ambient pressure. For a NEWTONian fluid, the shear stress tensor $\boldsymbol{S}$, with components $S_{ij}$, is related by the law $\boldsymbol{S} = 2\mu\boldsymbol{D}$ to the strain rate tensor $\boldsymbol{D}$, the components of which are $D_{ij} = \frac{1}{2}\,(\partial u_j/\partial x_i + \partial u_i/\partial x_j)$ $(i, j = 1, 2)$ with $u_1 = u$, $u_2 = v$, $x_1 = x$, $x_2 = y$.

During the motion, the velocity vector vanishes at all solid boundaries while the free surface corresponds to a stream surface on which the external gas exerts a negligible shear action. The free surface itself is in equilibrium in the presence of the internal and external pressures and of the effect of the surface tension.

Finally, if one introduces the dimensionless temperature difference $\theta(x, y) = [T(x, y) - T_\infty]/T_\infty$, it will be assumed that the temperature distribution at the bottom can be expressed by $\theta(x, -D) = \varepsilon f(x/d)$ with $\varepsilon \ll 1$ and $f$ a square summable function in the interval $-1 < x/d < 1$. Furthermore, for convenience and without any loss of generality, as far as the solution procedure is concerned, it is assumed that the free surface encounters the walls at a right angle.

## 2. Basic Equations and the Solution Procedure

The problem described in the preceding chapter is governed by the following system of partial differential equations and related boundary conditions. In particular, the momentum equation corresponds to the Boussinesq's approximation of the full Navier-Stokes' equation [10]. Therefore, if $\varrho$ and $\mu$ are evaluated at $T = T_\infty$

*a — Continuity equation*

$$\nabla \cdot \boldsymbol{u} = 0 , \tag{2.1}$$

*b — Momentum equation*

$$\varrho \boldsymbol{u} \cdot \nabla \boldsymbol{u} = -\nabla \Phi + \varrho\beta\theta T_\infty \boldsymbol{g} + \mu\Delta \boldsymbol{u} , \tag{2.2}$$

*c — Energy equation*

$$\boldsymbol{u} \cdot \nabla\theta = k\Delta\theta , \tag{2.3}$$

*d — Equilibrium at the free surface*

$$(S_{yy} - \varphi' S_{xy} - \Phi)_{y=\varphi(x)} = \{\sigma[\varphi'/(1 + \varphi'^2)^{1/2}]' - \varrho g\varphi\}_{y=\varphi(x)} , \tag{2.4}$$

with the *associated conditions*

$$\boldsymbol{u}(\pm d, y) = \boldsymbol{u}(x, -D) = 0 , \tag{2.5}$$

$$(v - \varphi' u)_{y=\varphi(x)} = 0 , \tag{2.6}$$

$$\{(S_{yy} - S_{xx})\,\varphi' + (1 - \varphi'^2)\,S_{yx}\}_{y=\varphi(x)} = 0 , \tag{2.7}$$

$$\{\pm \partial\theta/\partial x + (h_T/\lambda)\,\theta\}_{x=\pm d} = 0 , \tag{2.8}$$

$$\theta(x, -D) = \varepsilon f(x/d) , \tag{2.9}$$

$$(\partial\theta/\partial y - \varphi'\, \partial\theta/\partial x)_{y=\varphi(x)} = 0 , \tag{2.10}$$

$$\varphi'(\pm d) = 0 . \tag{2.11}$$

The symbols $\nabla$ and $\Delta$ indicate the gradient operator and the Laplace operator, respectively.

Moreover, the equation expressing the fact that the liquid volume within the trench has a constant value can be put in the form

$$\left(\frac{1}{2d}\right) \int_{-d}^{d} \varphi(x)\, \mathrm{d}x = 0 . \tag{2.12}$$

The system (2.5—2.12) will be solved by applying the so-called Lagrangean domain perturbation technique which was proposed by D. D. Joseph in a pioneer paper [11] followed by an application [12]. Without too many details, Joseph's method consists in assuming that the solution of the problem is known in a reference domain $D_0$. Then the solution in the actual domain $D_\varepsilon$ is expanded as a power series of some parameter $\varepsilon$. The coefficients of the series are the substancial derivatives of the unknowns, with respect to $\varepsilon$, and such derivatives are evaluated in $D_0$. The convergence of the series has not been generally proved from a mathematical point of view, except for some peculiar situations to which the present case can be related [13].

In the present circumstance, $D_\varepsilon$ corresponds to $D_\varepsilon = \{(x, y) \mid -d \leqq x \leqq d;\ -D \leqq y \leqq \varphi(x)\}$ and can be transformed into the reference domain $D_0 = \{(x_0, x_0) \mid -d \leqq x_0 \leqq d;\ -D \leqq y_0 \leqq 0\}$, wherein the fluid at rest is contained, by the univoquely invertible (analytic in $\varepsilon$) mapping $x = x_0$, $y = \chi(x_0, y_0; \varepsilon) = \varphi(x_0) + [D + \varphi(x_0)]\, y_0/D$. The unknown vector $\boldsymbol{q}(u, v, \theta, \Phi, \varphi) = \boldsymbol{q}(x, y; \varepsilon)$ is then expanded as

$$\boldsymbol{q}(x, y; \varepsilon) = \sum_{m=0}^{\infty} \boldsymbol{q}^{[m]}(x_0, y_0)\, \varepsilon^m/m! \tag{2.13}$$

where

$$\boldsymbol{q}^{[m]}(x_0, y_0) = \{\mathrm{d}^m \boldsymbol{q}(x_0, y_0)/\mathrm{d}\varepsilon^m\}_{\varepsilon=0} .$$

For a function $F(x, y; \varepsilon)$ the relations between the substancial derivatives $F^{[m]}$ and the partial derivatives $F^{\langle m\rangle}$ hold

$$F^{[0]} = F^{\langle 0\rangle} ,$$

$$F^{[1]} = F^{\langle 1\rangle} + \chi^{[1]}(\partial F^{\langle 0\rangle}/\partial y) ,$$

$$F^{[2]} = F^{\langle 2\rangle} + 2\chi^{[1]}(\partial F^{\langle 1\rangle}/\partial y) + \chi^{[2]}(\partial F^{\langle 0\rangle}/\partial y) + \chi^{[1]}(\partial^2 F^{\langle 0\rangle}/\partial y^2)$$

etc., where $\chi^{[m]} = \chi^{\langle m\rangle}$ and $\varphi^{[m]} = \varphi^{\langle m\rangle}$ as $x_0$, $y_0$ are independent of $\varepsilon$ and $\chi = \chi(x_0, y_0; \varepsilon)$.

After the problem is solved in $D_0$, the solution in $D_\varepsilon$ can be easily obtained by the inverse mapping. In dealing with the situation proposed herein, the series expansion will be truncated to $m = 1$, no major mathematical difficulties arising for higher values of $m$.

It is immediately realized that, for $m = 0$, the corresponding quantities $\{u^{\langle 0\rangle}, v^{\langle 0\rangle}, \theta^{\langle 0\rangle}, \Phi^{\langle 0\rangle}, \varphi^{\langle 0\rangle}\}$ are all equal to zero. Therefore, the first approximation to the system (2.5—2.12), to be solved in $D_0$, is obtained from the corresponding following set of equations

$$\nabla \cdot \boldsymbol{u}^{\langle 1\rangle} = 0 , \tag{2.14}$$

$$-\nabla \Phi^{\langle 1\rangle} + \varrho\beta\theta^{\langle 1\rangle} T_\infty \boldsymbol{g} + \mu\Delta \boldsymbol{u}^{\langle 1\rangle} = 0 , \tag{2.15}$$

$$\Delta\theta^{\langle 1\rangle} = 0 , \tag{2.16}$$

$$(2\mu\, \partial v^{\langle 1\rangle}/\partial y_0 - \Phi^{\langle 1\rangle})_{y_0=0} = \sigma\, \mathrm{d}^2\Phi^{\langle 1\rangle}/\mathrm{d}x_0^2 - \varrho g \Phi^{\langle 1\rangle} , \tag{2.17}$$

and the associated boundary conditions

$$u^{\langle 1\rangle}(\pm d, y_0) = u^{\langle 1\rangle}(x_0, -D) = 0\,, \tag{2.18}$$

$$(v^{\langle 1\rangle})_{y_0=0} = 0\,, \tag{2.19}$$

$$(S^{\langle 1\rangle}_{x_0y_0})_{y_0=0} = 0\,, \tag{2.20}$$

$$\{\pm \partial\theta^{\langle 1\rangle}/\partial x_0 + (h_T/\lambda)\,\theta^{\langle 1\rangle}\}_{x_0=\pm d} = 0\,, \tag{2.21}$$

$$\theta^{\langle 1\rangle}(x_0, -D)\,\varphi = f(x_0/d)\,, \tag{2.22}$$

$$(\partial\theta^{\langle 1\rangle}/\partial y_0)_{y_0=0} = 0\,, \tag{2.23}$$

$$(\mathrm{d}\varphi^{\langle 1\rangle}/\mathrm{d}x_0)_{x_0=\pm d} = 0 \tag{2.24}$$

with the integral relation

$$(1/2\,d)\int_{-d}^{d} \varphi^{\langle 1\rangle}(x_0)\,\mathrm{d}x_0 = 0\,. \tag{2.25}$$

From now on, the indexes related to the derivatives with respect to $\varepsilon$ will be omitted for convenience.

It is immediate to realize that the solution to the system (2.14—2.25) is greatly simplified by the fact that the equations in the unknowns $\theta$, $u$ and $\varphi$ can be solved in sequence.

### 3. The solution for the temperature and the flow field

It is suitable to express the various quantities in dimensionless form, and to introduce the stream function $\psi(x_0, y_0)$ defined by $u = \partial\psi/\partial y_0$ and $v = -\partial\psi/\partial x_0$. Let be $\xi = x_0/d$, $\eta = y_0/d$, and $\tilde{D} = D/d$. Let also be $\Psi(\xi,\eta) = (\mu/\varrho\beta T_\infty g d^3)\times \times\, \psi(x_0, y_0)$ and $B_i = h_T d/\lambda$, with $B_i$ indicating the BIOT modulus [9].

The equation (2.16) for the temperature distribution, with the conditions (2.21—2.23) corresponds to a heat conduction problem. To avoid lengthy mathematical formulae, it will be assumed here that the function $f(x_0/d)$ in equation (2.22), is the even function $g(\xi)$. There would be no difficulty in proceeding with an odd function.

Therefore, it can be easily proved [8] that the temperature distribution is given by

$$\theta(\xi,\eta) = 2\sum_{k=1}^{\infty} \frac{(B_i^2 + \alpha_k^2)\cos\alpha_k\xi\,\cosh\alpha_k\eta}{B_i(1+B_i) + \alpha_k^2\cosh\alpha_k}\int_0^1 g(\xi)\cos\alpha_k\xi\,\mathrm{d}\xi \tag{3.1}$$

with $\alpha_k$ the $k$-th root of $\alpha\,\mathrm{tg}\,\alpha = B_i$.

The set (2.14—2.15) with the boundary conditions (2.18—2.20) reduces to a classical STOKES problem in $\Psi(\xi,\eta)$:

$$\Delta^2\Psi = \partial\theta/\partial\xi\,, \tag{3.2}$$

$$\Psi(\pm 1,\eta) = (\partial\Psi/\partial\xi)_{\xi=\pm 1} = 0\,, \tag{3.3}$$

$$\Psi(\xi, 0) = (\partial^2\Psi/\partial\eta^2)_{\eta=0} = \Psi(\xi, -\tilde{D}) = (\partial\Psi/\partial\eta)_{\eta=-\tilde{D}} = 0\,. \tag{3.4}$$

The existence and uniqueness of the solution to the fluid motion problem can be straightforwardly proved from a functional point of view.

Let $R$ be the rectangle $\{(\xi,\eta) \mid -1<\xi<1,\ -\tilde{D}<\eta<0\}$ and let its boundary $\partial R$ be given by the lines $\gamma_1 = \{(1,\eta) \mid -\tilde{D} \leqq \eta \leqq 0\}$; $\gamma_2 = \{(-1,\eta) \mid -\tilde{D} \leqq \eta \leqq 0\}$; $\gamma_3 = \{(\xi, -\tilde{D}) \mid -1 \leqq \xi \leqq 1\}$; $\gamma_4 = \{(\xi, 0) \mid -1 \leqq \leqq \xi \leqq 1\}$. Furthermore, let $\bar{R}$ be the closure of $R$.

The proposed problem corresponds to solving the differential system

$$\Delta^2\Psi(\xi,\eta) = r(\xi,\eta) \qquad \forall\ (\xi,\eta) \in R\,,$$

$$\Psi|_{\partial R} = 0\,,$$

$$\left.\frac{\partial\Psi}{\partial n}\right|_{\gamma_1\cup\gamma_2\cup\gamma_3} = 0; \quad \left.\frac{\partial^2\Psi}{\partial\eta^2}\right|_{\gamma_4} = 0$$

where $n$ stays for the normal to $\gamma_1 \cup \gamma_2 \cup \gamma_3$, pointing outward.

If one considers the HILBERT's energy space $\mathcal{H}$ which completes the space

$$\left\{\Psi \in C^{(4)}(\bar{R})\ \middle|\ \Psi|_{\partial R} = 0;\ \left.\frac{\partial\Psi}{\partial n}\right|_{\gamma_1\cup\gamma_2\cup\gamma_3} = 0;\ \left.\frac{\partial^2\Psi}{\partial\eta^2}\right|_{\gamma_4} = 0\right\}$$

with respect to the norm $\|\Psi\|^2 = \iint_R [\Delta\Psi(\xi,\eta)]^2\,\mathrm{d}\xi\,\mathrm{d}\eta$, then the problem has a unique solution in $\mathcal{H}$ [14], provided that $r(\xi,\eta) \in \mathcal{L}^{(2)}(R)$. This can be shown by noting that the solution corresponds to minimizing, in $\mathcal{H}$, the quadratic functional (or energy functional)

$$J(\Psi) = \iint_R [\Delta\Psi(\xi,\eta)]^2\,\mathrm{d}\xi\,\mathrm{d}\eta - 2\iint_R r(\xi,\eta)\,\Psi(\xi,\eta)\,\mathrm{d}\xi\,\mathrm{d}\eta\,.$$

In fact, the solution to the differential problem (3.2—3.4) corresponds to the unique element $\Psi$ which minimizes the quadratic, energy functional $J(\Psi)$.

If one puts $\Psi(\xi, \eta) = \Psi^*(\xi, \eta) + G(\xi, \eta)$, with $\Psi^*$ a particular integral of (3.2) and $G(\xi, \eta)$ the general integral of the homogeneous equation associated to the (3.2), then one obtains

$$\Psi^*(\xi, \eta) = (\tfrac{1}{8}) \sum_{k=1}^{\infty} (B_k/\alpha_k b_k) \{b_k(\xi^2 - 1) \sin \alpha_k\xi - B_i^2\alpha_k \cos \alpha_k\xi + B_i\alpha_k^2 \sin \alpha_k\xi\} \cosh \alpha_k\eta \tag{3.5}$$

where $B_k = 2(B_i^2 + \alpha_k^2)/\{[B_i(1 + B_i) + \alpha_k^2] \cosh \alpha_k\tilde{D}\} \int_0^1 g(\xi) \cos \alpha_k\xi \, d\xi$ and $b_k = (\alpha_k^2/2) [B_i(1 - B_i) - \alpha_k^2]$.

$G(\xi, \eta)$ is given by the solution of the following system

$$\Delta^2 G = 0 , \tag{3.6}$$

$$G(\pm 1, \eta) = (\partial G/\partial\xi)_{\xi=\pm 1} = 0 , \tag{3.7}$$

$$G(\xi, 0) = -\tfrac{1}{8} \sum_{k=1}^{\infty} B_k l_k(\xi)/b_k\alpha_k , \tag{3.8}$$

$$(\partial^2 G/\partial\eta^2)_{\eta=0} = -\tfrac{1}{8} \sum_{k=1}^{\infty} B_k\alpha_k l_k(\xi)/b_k , \tag{3.9}$$

$$G(\xi, -\tilde{D}) = -\tfrac{1}{8} \sum_{k=1}^{\infty} B_k(\cosh \alpha_k\tilde{D}) \, l_k(\xi)/\alpha_k b_k , \tag{3.10}$$

$$(\partial G/\partial\eta)_{\eta=-\tilde{D}} = \tfrac{1}{8} \sum_{k=1}^{\infty} B_k(\sinh \alpha_k\tilde{D}) \, l_k(\xi)/b_k \tag{3.11}$$

where $l_k(\xi) = b_k(\xi^2 - 1) \sin \alpha_k\xi - \alpha_k B_i^2\xi \cos \alpha_k\xi + B_i\alpha_k^2 \sin \alpha_k\xi$.

To the bi-harmonic equation (3.6) it can be given a solution in terms of a double series of the PAPKOVICH-FADLE's type

$$G(\xi, \eta) = \sum_{\substack{n=-\infty \\ n\neq 0}}^{\infty} (C_n \, e^{s_n\eta} + D_n \, e^{-s_n\eta}) \, \Xi^{(n)}(\xi)/s_n^2 \tag{3.12}$$

where

$$\Xi^{(n)} = s_n(\cos s_n \sin s_n\xi - \xi \sin s_n \cos s_n\xi) \tag{3.13}$$

with $s_n$ the complex roots of $\sin 2s = 2s$, in the first quadrant, and such that the conjugate $\bar{s}_n = s_{-n}$.

The original problem has therefore been reduced to the determination of the coefficients $C_n$ and $D_n$ appearing in the proposed series (3.12). This can be done by exploiting the bi-orthogonality properties of $\Sigma(\xi, \eta)$ and by applying a method proposed by SMITH [15] for the solution of an elasticity problem. The infinite set of algebraic equations in $C_n$ and $D_n$, after some manipulation, is then

$$C_n + D_n = (1/K_n) \int_{-1}^{1} [t_1^{(n)}, t_2^{(n)}] \cdot \begin{vmatrix} 0 & -1 \\ 1 & 2 \end{vmatrix} \cdot \begin{bmatrix} (\partial^2 G/\partial\eta^2)_{\eta=0} \\ (\partial^2 G/\partial\xi^2)_{\eta=0} \end{bmatrix} d\xi , \tag{3.14}$$

$$K_n(C_n \, e^{-s_n\tilde{D}} + D_n \, e^{s_n\tilde{D}}) + \sum_{\substack{m=-\infty \\ m\neq 0}}^{\infty} A_{mn}\left(C_m \, e^{-s_m\tilde{D}} - \frac{1+s_m}{1-s_m} D_m \, e^{s_m\tilde{D}}\right) - \int_{-1}^{1} [t_1^{(n)}, t_2^{(n)}] \cdot \begin{vmatrix} 0 & -1 \\ 1 & 2 \end{vmatrix} \times$$
$$\times \begin{bmatrix} (\partial G/\partial\eta)_{\eta=-\tilde{D}} \\ (\partial^2 G/\partial\xi^2)_{\eta=-\tilde{D}} \end{bmatrix} d\xi \tag{3.15}$$

where $K_n = -4 \sin^4 s_n$ and

$$t_1^{(n)}(\xi) = s_n(\cos s_n + 2 \sin s_n) \cos s_n\xi - s_n\xi \cos s_n \sin s_n\xi ,$$

$$t_2^{(n)}(\xi) = \Xi^{(n)}(\xi) ,$$

$$A_{mn} = \frac{1 - s_m}{s_m} \int_{-1}^{1} \Xi^{(m)}(\xi) \, t_1^{(n)}(\xi) \, d\xi .$$

In a completely analogous way one can obtain the analytic solution to the case of a bottom temperature distribution which is an odd function of $\xi$.

Showing the convergence of a series like (3.12) represents a quite difficult task from an analytic point of view. It will be demonstrated in a further chapter how good are the convergence characteristics of (3.12) from a numerical point of view, in a particularly significant case. This enables one to limit the set of algebraic equations (3.14—3.15) to a reasonable number of terms.

### 4. The Shape of the Free Surface

Let the dimensionless expression of the reduced pressure be given by $P(\xi, \eta) = \Phi(x_0, y_0)/\varrho\beta T_\infty gd$ while for the free boundary equation let $h(\xi) = \sigma\varphi(x_0)/\varrho\beta T_\infty gd^3$. The equation (2.17) can then be written in the form

$$d^2h(\xi)/d\xi^2 - N_\sigma h(\xi) = -P(\xi, 0) - 2 [\partial^2\Psi(\xi, \eta)/\partial\xi \, \partial\eta]_{\eta=0} \tag{4.1}$$

where $N_\sigma$ is a non-dimensional parameter equal to $\varrho g d^2/\sigma$. The expression for $P(\xi, 0)$ in equation (4.1) can be obtained by immediate integration of the $\eta$-component of the equation (2.15), which can be written as

$$\partial P(\xi,\eta)/\partial\eta = \theta(\xi,\eta) - \Delta[\partial\Psi(\xi,\eta)/\partial\xi] . \tag{4.2}$$

The relations (2.24—2.25) now assume the form

$$[\mathrm{d}h/\mathrm{d}\xi]_{\xi=\pm1} = 0; \qquad \int_{-1}^{1} h(\xi)\,\mathrm{d}\xi = 0 . \tag{4.3}$$

From (4.1—4.3) the analytic solution for the first approximation of the free boundary is obtained

$$h(\xi) = \sum_{\substack{n=-\infty\\ n\neq 0}}^{\infty} \left\{(a_1 \cos s_n\xi + a_2\xi \sin s_n\xi) + 2a_3 \cosh s_n\xi\right\} + a_4$$

where

$$a_1 = -2(C_n - D_n)\,(2\sin s_n - s_n \cos s_n)/(s_n^2 + N_\sigma) - 2s_n^2(C_n - D_n) \sin s_n/(s_n^2 + N_\sigma)^2 ,$$

$$a_2 = -2(C_n - D_n)\, s_n \sin s_n/(s_n^2 + N_\sigma) ,$$

$$a_3 = -(1/2N_\sigma \sinh N_\sigma) \sum_{n=-\infty}^{\infty} [(a_2 - a_1 s_n) \sin s_n + a_2 s_n \cos s_n] ,$$

$$a_4 = -(2a_3 \sinh N_\sigma/N_\sigma) - \sum_{\substack{n=-\infty\\ n\neq 0}}^{\infty} [(a_1/s_n + a_2/s_n^2) \sin s_n - (a_2/s_n) \cos s_n] .$$

## 5. Application

In order to show numerically the convergence characteristics of the proposed analytic solution and to investigate some basic features of the natural convection in a trench, a meaningful situation will now be investigated. In the following, it will be assumed that the bottom of the open channel is kept at the constant temperature $T_b$, with $(T_b - T_\infty)/T_\infty = \varepsilon \ll 1$, so that $g(\xi) = 1$ in the equation (3.1). It will also be taken $B_i = 1$ as an indicative value.

Table 1 and Table 2 show the characteristics of the convergence of the series (3.12) to the function (3.5) for two different values of the depth $\tilde{D}$ (one should recall that $\Psi(\xi,\eta) = \Psi^*(\xi,\eta) + G(\xi,\eta)$). In particular, Table 1 shows, for $\tilde{D} = 0.5$, the computed values of $f_1(\xi) = -\Psi^*(\xi, 0)$ and $g_1(\xi) = -\Psi^*(\xi, -0.5)$ and their series representations $G(\xi, 0)|_N$ and $G(\xi, -0.5)|_N$, respectively, at selected values of $\xi$. $N$ stays for the number of terms taken into account. Table 2, for $\tilde{D} = 2$, gives $f_2(\xi) = -\Psi^*(\xi, 0)$ and $g_2(\xi) = -\Psi^*(\xi, -2)$ and their series expansions $G(\xi, 0)|_N$ and $G(\xi, -2)|_N$, respectively. As one can note, in both cases the convergence is excellent, and improves at increasing values of $\tilde{D}$, $N = 3$ being sufficient to insure a truncation error of few percents for the case of Table 1, while in the case of Table 2 it is sufficient to take $N = 2$.

The essential characteristics of the temperature distribution and of the flow structure are given in Figs. 1 through 6. In all the cases, $B_i = 1$, and the bottom temperature is constant. Fig. 1 shows some selected isotherms in the case of a square cavity. Thermal gradients in the $\eta$-direction prevail near the bottom surface, while $\xi$-gradient become significant near the solid walls and in the upper region. The isotherms, as the streamlines, are symmetric with respect to $\xi$.

As a consequence of the temperature distribution of Fig. 1, the streamlines of Fig. 2 are obtained which correspond to a vortex in the reference domain. The traces on the $(\xi,\eta)$-plane of the free surface are given in Fig. 3 for various values of the dimensionless number $N_\sigma$. As $N_\sigma$ increases, the free surface tends to the plane $\eta = 0$, due to decreasing importance of the surface tension. In all cases one notes that the absolute maximum of the boundary upper surface does not stay on the $\eta$ axis, due to the counterclockwise circulation in the represented region. After mapping the reference domain back to the actual domain, one has the flow structure of Fig. 4, which is not to scale in the $\eta$-direction, as the vortex deformation could not otherwise be represented in a small drawing.

Fig. 5, for $\tilde{D} = 0.5$, and Fig. 6, for $\tilde{D} = 1$, are eventually given to provide a comparison of the influence of the geometric aspect ratio of the trench on the distribution of the streamlines.

Table 1

| $\xi$ | $f_1(\xi)$ | $G(\xi, 0)\vert_N$ | | | | | | |
|---|---|---|---|---|---|---|---|---|
| | | $N = 1$ | $N = 2$ | $N = 3$ | $N = 4$ | $N = 5$ | $N = 6$ | $N = 7$ |
| 0.2 | 0.000092 | 0.000069 | 0.000096 | 0.000091 | 0.000092 | 0.000091 | 0.000092 | 0.000092 |
| 0.4 | 0.000028 | -0.000018 | 0.000033 | 0.000028 | 0.000026 | 0.000028 | 0.000028 | 0.000027 |
| 0.6 | −0.000143 | −0.000164 | −0.000155 | −0.000141 | −0.000141 | −0.000144 | −0.000144 | −0.000144 |
| 0.8 | −0.000160 | −0.000134 | −0.000163 | −0.000165 | −0.000162 | −0.000160 | −0.000160 | −0.000159 |

| $\xi$ | $g_1(\xi)$ | $G(\xi, -0.5)\vert_N$ | | | | | | |
|---|---|---|---|---|---|---|---|---|
| | | $N = 1$ | $N = 2$ | $N = 3$ | $N = 4$ | $N = 5$ | $N = 6$ | $N = 7$ |
| 0.2 | 0.001920 | 0.002120 | 0.001916 | 0.001916 | 0.001920 | 0.001921 | 0.001920 | 0.001920 |
| 0.4 | 0.002940 | 0.003115 | 0.002932 | 0.002943 | 0.002941 | 0.002939 | 0.002941 | 0.002940 |
| 0.6 | 0.002560 | 0.002506 | 0.002587 | 0.002560 | 0.002558 | 0.002560 | 0.002560 | 0.002560 |
| 0.8 | 0.001079 | 0.000926 | 0.001058 | 0.001079 | 0.001084 | 0.001083 | 0.001081 | 0.001080 |

Table 2

| $\xi$ | $f_2(\xi)$ | $G(\xi, 0)\|_N$ | | | | | | |
|---|---|---|---|---|---|---|---|---|
| | | $N = 1$ | $N = 2$ | $N = 3$ | $N = 4$ | $N = 5$ | $N = 6$ | $N = 7$ |
| 0.2 | 0.000451 | 0.000465 | 0.000450 | 0.000451 | 0.000451 | 0.000451 | 0.000451 | 0.000451 |
| 0.4 | 0.000687 | 0.000707 | 0.000685 | 0.000688 | 0.000688 | 0.000687 | 0.000687 | 0.000688 |
| 0.6 | 0.000594 | 0.000599 | 0.000597 | 0.000593 | 0.000593 | 0.000594 | 0.000594 | 0.000594 |
| 0.8 | 0.000248 | 0.000236 | 0.000250 | 0.000250 | 0.000248 | 0.000248 | 0.000247 | 0.000247 |
| $\xi$ | $g_2(\xi)$ | $G(\xi, -2)\|_N$ | | | | | | |
| | | $N = 1$ | $N = 2$ | $N = 3$ | $N = 4$ | $N = 5$ | $N = 6$ | $N = 7$ |
| 0.2 | 0.001920 | 0.002170 | 0.001917 | 0.001916 | 0.001920 | 0.001921 | 0.001920 | 0.001920 |
| 0.4 | 0.002940 | 0.003151 | 0.002931 | 0.002944 | 0.002941 | 0.002939 | 0.002941 | 0.002940 |
| 0.6 | 0.002560 | 0.002488 | 0.002588 | 0.002560 | 0.002558 | 0.002560 | 0.002560 | 0.002560 |
| 0.8 | 0.001079 | 0.000897 | 0.001056 | 0.001078 | 0.001084 | 0.001083 | 0.001081 | 0.001080 |

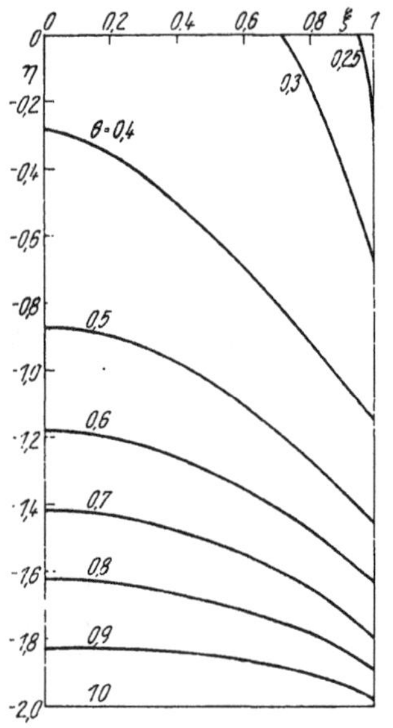


Fig. 1. Isotherms distribution in the reference domain, $\tilde{D} = 2$

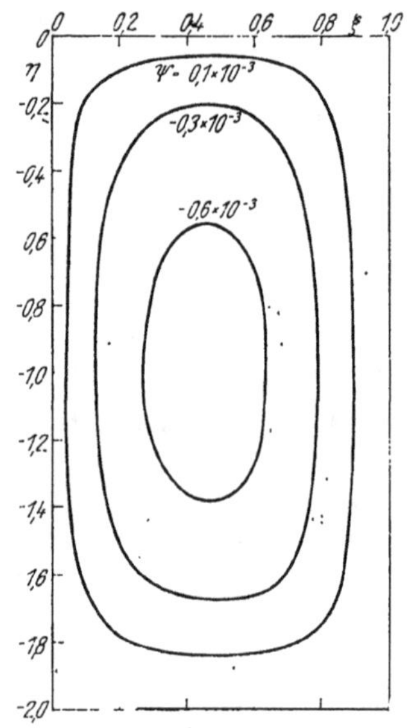


Fig. 2. Streamlines distribution in the reference domain, $\tilde{D} = 2$

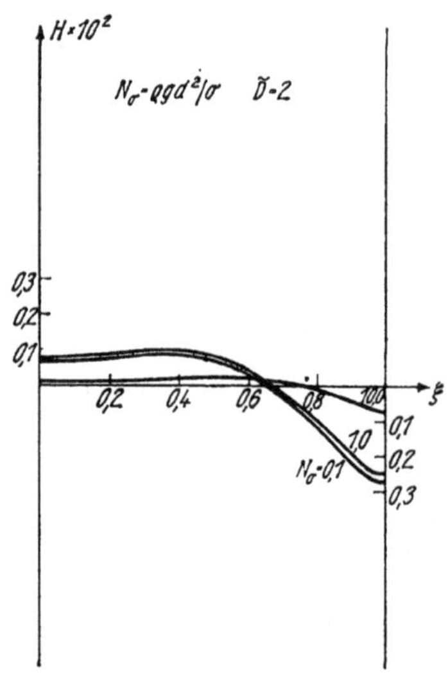


Fig. 3. Free surface shapes at various $N_\sigma$; $\tilde{D} = 2$

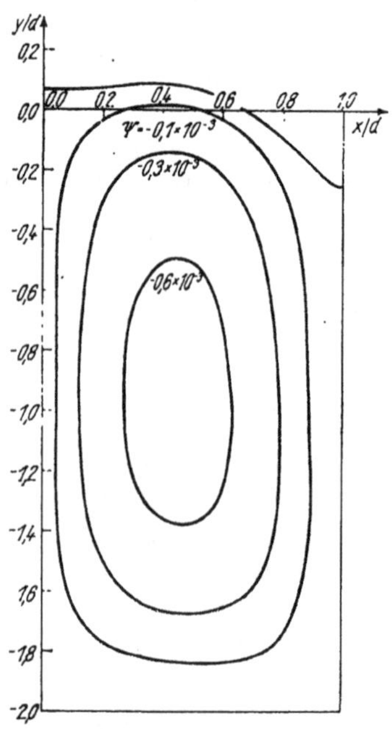


Fig. 4. Streamlines distribution in the actual domain (not to scale), $\tilde{D} = 2$

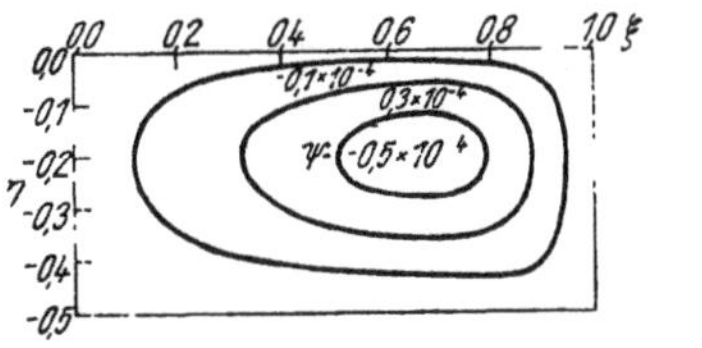


Fig. 5. Streamlines distribution, $\tilde{D} = 0.5$

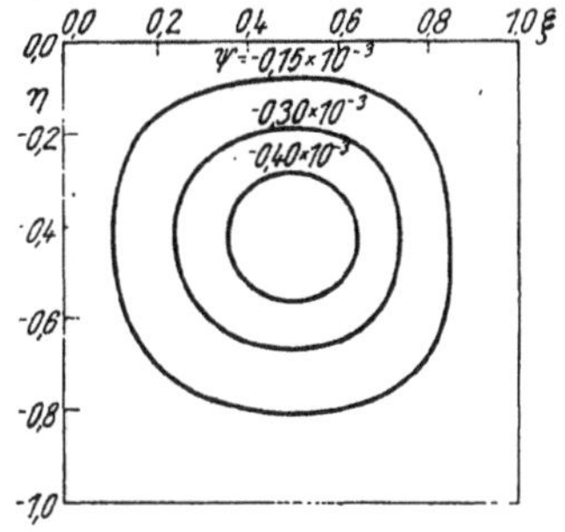


Fig. 6. Streamlines distribution, $\tilde{D} = 1$

## Acknowledgement

This work was partially supported by the Italian National Research Council (G.N.F.M. and G.N.A.F.A and contract N. 79.02817.11).

## References

1 Batchelor, G. K., Heat Transfer by Free Convection across a Closed Cavity between Vertical Boundaries at Different Temperatures, Q. Appl. Math. **12**, 209–233 (1954).
2 Lighthill, M. J., Theoretical Considerations of Free Convection in Tubes, Q. J. Mech. Appl. Math. **6**, 398–439 (1953).
3 Ostrach, S., Natural Convection in Enclosures, in "Advances in Heat Transfer", vol. 8, Irvine and Hartnett eds., Acad. Press, New York 1972.
4 Japikse, D., Advances in Thermosiphon Technology, in "Advances in Heat Transfer", vol. 9, Irvine and Hartnett eds., Acad. Press, New York 1973.
5 Yin, S. H., Wung, T. Y. and Chen, K., Natural Convection in an Air Layer Enclosed within Rectangular Cavities, Int. J. Heat Mass Transfer **21**, 307–315 (1978).
6 Morrison, G. L. and Tran, V. Q., Laminar Flow Structure in Vertical Free Convective Cavities, Int. J. Heat Mass Transfer **21**, 203–213 (1978).
7 Mallinson, G. D. and De Vahl Davis, G., Three Dimensional Natural Convection in a Box: a Numerical Study, J. Fluid Mech. **83**, 1–31 (1977).
8 Carslaw, H. S., and Jaeger, J. C., Conduction of Heat in Solids, Oxford University Press, 1959.
9 Arpaci, V., Conduction Heat Transfer, Addison Wesley Publ. Co. Reading 1966.
10 Joseph, D. D., Stability of Fluid Motions, vol. 2, Springer Verlag, Berlin 1976.
11 Joseph, D. D., Domain Perturbation: the Higher Order Theory of Infinitesimal Water Waves, Arch. Rat. Mech. An. **51**, 295–303 (1973).
12 Joseph, D. D. and Sturges, L., The Free Surface on a Liquid Filling a Trench Heated from its Side, J. Fluid Mech. **69**, 565–589 (1975).
13 Sattinger, D. H., On the Free Surface of a Viscous Fluid Motion, Proc. Royal Soc. London A, **349**, 183–204 (1976).
14 Mikhlin, S. G., Mathematical Physics, an Advanced Course, North-Holland Publ. Co., Amsterdam 1970.
15 Smith, R. C., The Bending of a Semi-Infinite Strip, Aus. J. Sc. Res. **5**, 227–237 (1952).

*Anschriften*: Prof. Dr. Luciano M. de Socio, Istituto di Meccanica Applicata, Aerodinamica e Gasdinamica, Politecnico di Torino, Corso Duca degli Abruzzi 24, 10127 TORINO, ITALY — Dr. Luciano Misici, Istituto di Matematica, Università di Camerino, Via V. Venanzi 16, 62032 CAMERINO (MC), ITALY

ZAMM 60, 39 – 48 (1980)

P. D. ARIEL/B. D. AGGARWALA

# RAYLEIGH-TAYLOR Instability of a Visco-Elastic Fluid

*Es wird der Charakter des Gleichgewichtes einer inkompressiblen viskoelastischen Flüssigkeit mit veränderlicher Dichte in der vertikalen Richtung untersucht. Wie gezeigt wird, ist die Lösung durch ein Variationsproblem bestimmt. Von diesem Prinzip wird Gebrauch gemacht, um Näherungslösungen für eine Oldroyd-Flüssigkeit mit exponentiell veränderlicher Dichte zu erhalten, die durch zwei horizontale Ränder begrenzt wird. Es wird gezeigt, daß die Rate, mit der sich das System vom Gleichgewichtszustand entfernt, mit der Spannungs-Relaxationszeit anwächst und fällt, wenn die Dehnungs-Verzögerungszeit steigt. Es wird gefunden, daß bei einer Konfiguration, wo Gravitationseffekte zu vernachlässigen sind, Wellen durch gewisse Störungen erregt werden können, bei denen in einer Newtonschen Flüssigkeit keine Oszillationen möglich sind. Die Rotationseffekte werden ebenfalls behandelt. Schließlich wird die Brauchbarkeit des Variationsprinzips durch Vergleich einiger exakter Ergebnisse mit den Näherungslösungen überprüft.*

*The character of equilibrium of an incompressible visco-elastic fluid of variable density stratified in the vertical direction has been investigated. It is demonstrated that the solution is characterized by a variational principle. Use has been made of this principle to obtain approximate solution for an Oldroyd fluid of exponentially varying density confined between two horizontal boundaries. It is shown that the rate at which the system departs from equilibrium increases with the stress relaxation time and decreases as the strain retardation time increases. For a configuration in which gravity effects are ignored, it is found that the waves can be excited for some disturbances for which no oscillatory motion is possible in a newtonian fluid. The effects of rotation are also considered. Finally the usefulness of variational principle is checked by comparing some exact results with approximate ones.*

Исследуется характер равновесия несжимаемой вязко-упругой жидкости переменной плотности, расслоенной в вертикальном направлении. Показано, что решение характеризуется вариационным принципом. Этот принцип используется для подсчёта приближённого решения для жидкости Олдройда с экспоненциально изменяющейся плотностью, находящейся между двумя горизонтальными границами. Показано, что скорость, с которой система отклоняется от равновесия, возрастает со временем релаксации напряжения и убывает с возрастанием времени запаздывания деформации. Для случая, когда гравитационные эффекты не учитываются, обмаружено, что волны могут возбуждаться при некоторых возмущениях, при которых колебательное движение в ньютоновской жидкости не возможно. Рассмотрены также эффекты вращения. В заключение выясняется преимущество вариационного принципа путём сравнения некоторых точных результатов с приближёнными.

## Introduction

Since the pioneer work of RAYLEIGH [1], the stability of a fluid of variable density has commanded the attention of numerous researchers. RAYLEIGH, in his classical paper derived the following fundamental result: *A fluid of variable density is stable or unstable according as its density decreases everywhere or increases anywhere in the upward direction.* He also obtained explicit solutions for two arrangements of fluid (i) one fluid of uniform density topped by another fluid of different uniform density (ii) fluid having exponentially varying density confined between two horizontal planes.

For a real fluid exhibiting viscosity, this problem was first investigated by HARRISON [2] who obtained a power series for the characteristic values valid in the limit of zero viscosity. CHANDRASEKHAR [3] reconsidered the aforementioned problem and gave a detailed and more complete treatment. Besides he demonstrated that the solution was characterized by a variational principle. HIDE [4] made use of this variational principle to get simple though approximate solutions for the density distributions mentioned above.

Of late, considerable interest has been aroused in the investigation of instabilities of another class of real fluids, namely, the non-newtonian fluids. Even though, several kinds of instabilities have been considered, it seems, not enough attention has been paid to the RAYLEIGH-TAYLOR instability.

In the present paper, it is our endeavor to examine the RAYLEIGH-TAYLOR instability of one of the most common non-newtonian fluids, the OLDROYD fluid.

## Formulation of the Problem

Consider a visco-elastic incompressible fluid of density $\varrho$, depending on the vertical coordinate $z$ only, stratified in the vertical direction. For the model proposed by OLDROYD [5], the relation between $\tau_{ij}$, the deviatoric stress tensor and $e_{ij}$ the rate of strain tensor is

$$\left(1 + \lambda_1 \frac{\mathrm{d}}{\mathrm{d}t}\right)\tau_{ij} = 2\eta_0 \left(1 + \lambda_2 \frac{\mathrm{d}}{\mathrm{d}t}\right) e_{ij}\,, \tag{1}$$

where $\lambda_1$ and $\lambda_2 (\lambda_1 > \lambda_2)$ are respectively the stress relaxation and strain retardation times, $\eta_0$ is the viscosity of the medium and $e_{ij}$ is given by

$$e_{ij} = \frac{1}{2}\left(\frac{\partial u_i}{\partial x_j} + \frac{\partial u_j}{\partial x_i}\right). \tag{2}$$

In equation (2), $u_i$ denotes the velocity at the point $x_i$.

The equation of motion is

$$\varrho \frac{\mathrm{d}u_i}{\mathrm{d}t} = -\frac{\partial p}{\partial x_i} + \frac{\partial}{\partial x_j}\tau_{ij} - g\varrho e_i , \tag{3}$$

where $p$ denotes the scalar pressure, $g$ is the acceleration due to gravity and $e_i$ ($= 0, 0, 1$) is a unit vector in the vertically upward direction.[1])

The equation for incompressibility of the medium is

$$\frac{\partial \varrho}{\partial t} + u_j \frac{\partial \varrho}{\partial x_j} = 0 \tag{4}$$

whence the equation of continuity can be written as

$$\frac{\partial u_j}{\partial x_j} = 0 . \tag{5}$$

In the equilibrium position the state is characterized by $u_i = 0$. Give a small disturbance to the system which produces a velocity field $u_i$ (also denoted as $(u, v, w)$). Associated with this perturbation are the corresponding perturbations $\delta\varrho$, $\delta\eta_0$, $\delta p$ in $\varrho$, $\eta_0$ and $p$ respectively.The linearized perturbation equations are

$$\varrho \frac{\partial u_i}{\partial t} = -\frac{\partial}{\partial x_i}\delta p + \frac{\partial}{\partial x_j}\tau_{ij} - g\delta\varrho e_i \tag{6}$$

$$\frac{\partial}{\partial t}\delta\varrho = -w\frac{\mathrm{d}\varrho}{\mathrm{d}z} \tag{7}$$

$$\frac{\partial u_j}{\partial x_j} = 0 . \tag{8}$$

Analysing the disturbance in normal modes, we seek solutions of equations (6) — (8) for which $x$, $y$, $t$ dependence is given by

$$\exp\,(ik_x x + ik_y y + nt) \tag{9}$$

where $k_x$ and $k_y$ are the horizontal components of the wave-vector $\boldsymbol{k}$ and $n$ is the growth rate of disturbance.

Making use of equations (1) and (2), one can rewrite the equations (6) to (8) as follows

$$n\varrho u = -ik_x\delta p + \mu(D^2 - k^2)\,u + D\mu(Du + ik_x w) \tag{10}$$

$$n\varrho v = -ik_x\delta p + \mu(D^2 - k^2)\,v + D\mu(Dv + ik_y w) \tag{11}$$

$$n\varrho w = -D\delta p + \mu(D^2 - k^2)\,w + 2D\mu\,Dw - g\delta\varrho \tag{12}$$

$$n\delta\varrho = -wD\varrho \tag{13}$$

$$ik_x u + ik_y v + Dw = 0 \tag{14}$$

where $D$ stands for d/d$z$, and

$$\mu = \eta_0 \frac{1 + \lambda_2 n}{1 + \lambda_1 n}, \tag{15}$$

$\mu$ may be termed as the modified coefficient of viscosity. Eliminating some of the perturbed quantities from equa tions (10) — (14), we obtain the following differential equation in $w$

$$n\{k^2\varrho w - D(\varrho Dw)\} - \frac{gk^2}{n}(D\varrho)w + \mu(D^2 - k^2)^2\,w + 2D\mu(D^2 - k^2)\,Dw + D^2\mu(D^2 + k^2)\,w = 0 . \tag{16}$$

To obtain an eigenvalue relation between $n$ and $k$, one must solve equation (16) subject to appropriate boundary conditions.

### Boundary Conditions

At the boundary, the fluid can not have a vertical component of velocity [1]. Thus

$$w = 0 \quad \text{on a boundary rigid or free.} \tag{17}$$

Further if the boundary is rigid, motion is prohibited on it, on account of viscosity which requires $u = v = w = 0$ on a rigid boundary, and by virtue of equation (14), it follows that

$$Dw = 0 \quad \text{on a rigid boundary}. \tag{18}$$

On the free boundary, the tangential stresses must vanish, i.e., $\tau_{xz} = \tau_{yz} = 0$. Now $\tau_{xz} = \mu(Du + ik_x w)$ and $\tau_{yz} = \mu(Dv + ik_y w)$. Therefore, $-ik_x\tau_{xz} - ik_y\tau_{yz} = \mu(D^2 + k^2)\,w$. Since $w = 0$, we have

$$D^2 w = 0 \quad \text{on a free boundary}. \tag{19}$$

[1]) The differentiation d/d$t$ is to be understood as a "material" one in (3), but as an "observer-invariant" one in (1). After linearisation, this difference disappears.

## Variational Principle

Multiplying equation (16) for $w_i$ by $w_j$ and integrating across the vertical extent of the fluid (denoted by $L$), we obtain after a number of integrations by part

$$n_i \int_L \varrho(Dw_i\, Dw_j + k^2 w_i w_j)\, dz - \frac{gk^2}{n_i} \int_L D\varrho\, w_i w_j\, dz + \int_L \mu[(D^2 + k^2)\, w_i (D^2 + k^2)\, w_j + 4k^2\, Dw_i\, Dw_j]\, dz = 0 \tag{20}$$

the integrated parts vanishing because of boundary conditions (17) and (18) or (19). Setting $i = j$ in equation (20), we get

$$n \int_L \varrho[(Dw)^2 + k^2 w^2]\, dz - \frac{gk^2}{n} \int_L D\varrho w^2\, dz + \int_L \mu\{[(D^2 + k^2)\, w]^2 + 4k^2 (Dw)^2\} dz = 0\,. \tag{21}$$

Consider a small change $\delta w$ in $w$ compatible with the boundary conditions of the problem. The corresponding increment $\delta n$ in $n$ can be obtained from equation (21). We have to the first order

$$-\frac{1}{2}\,\delta n \left( \int_L \varrho[(Dw)^2 + k^2 w^2]\, dz + \frac{gk^2}{n^2} \int_L D\varrho w^2\, dz + \int \frac{\partial u}{\partial n} \mu \{[(D^2 + k^2)\, w]^2 + 4k^2 (Dw)^2\}\, dz \right) =$$

$$= \int_L \left\{ n[k^2 \varrho w - D(\varrho Dw)] - \frac{gk^2}{n} (D\varrho)\, w + \mu(D^2 - k^2)^2 w + 2D\mu(D^2 - k^2)\, Dw + D^2\mu(D^2 + k^2)\, w \right\} \delta w\, dz\,. \tag{22}$$

From equation (22), we observe that a necessary and sufficient condition for $\delta n$ to be zero to the first order of smallness is that $w$ be the solution of the characteristic equation (16). Thus the present problem is characterized by a variational principle. In the next section use has been made of this fact to obtain an approximate solution for the case of a fluid of finite depth and exponentially varying density.

## The Case of a Continuously Stratified Fluid of Finite Depth

Consider a fluid layer confined between the planes $z = 0$ and $z = d$ for which $\varrho$ and $\eta_0$ have the following variations with respect to $z$

$$\varrho = \varrho_0 \exp \beta z \tag{23}$$

and

$$\eta_0 = \nu \varrho_0 \exp \beta z \tag{24}$$

where $\varrho_0$, $\nu$ and $\beta$ are constants.

Let us assume the following trial function for $w(z)$

$$w(z) = W \sin lz\,. \tag{25}$$

For $w$ to vanish at the boundary $z = d$, we require

$$l = \frac{\pi s}{d} \tag{26}$$

where $s$ is an integer.

Substituting for $w(z)$ in equation (21) and evaluating the integrals, we obtain the following dispersion relation in $n$ and $k$

$$n^2\left(l^2 + k^2 + \frac{1}{2}\beta^2\right) + \frac{\nu n(1 + \lambda_2 n)}{1 + \lambda_1 n} [(k^2 + l^2)^2 + 2\beta^2 l^2] - g\beta k^2 = 0\,. \tag{27}$$

Equation (27) reduces to the corresponding dispersion relation obtained by Hide [4] for a newtonian fluid ($\lambda_1 = \lambda_2 = 0$) under the assumption $|\beta d| \ll 1$. To non-dimensionalise equation (27), we measure $n$ and $k$ in terms of $(\pi^2 s^2 \nu/d^2)$ sec$^{-1}$ and $(\pi s/d)$ cm$^{-1}$ respectively. This gives

$$n^2\left(k^2 + 1 + \frac{1}{2}a^2\right) + \frac{n(1 + T_2 n)}{1 + T_1 n} [(k^2 + 1)^2 + 2a^2 k^2] - Gk^2 = 0 \tag{28}$$

where

$$T_1 = \frac{\lambda_1 \nu \pi^2 s^2}{d^2}\,, \quad T_2 = \frac{\lambda_2 \nu \pi^2 s^2}{d^2}\,, \tag{29), (30}$$

$$G = \frac{g\beta d^4}{\pi^4 s^4 \nu^2}\,, \quad a = \frac{\beta d}{\pi s}\,. \tag{31), (32}$$

Here $G$ has the form of Grashoff number and $T_1$ and $T_2$ can be reckoned as non-dimensional stress relaxation and retardation times respectively, and $a$ is a measure of variation of density.

If we make the assumption

$$|a| \ll 1 \tag{33}$$

which was also made by Hide, we shall reduce the number of parameters without essentially changing the physical setting of the problem. This assumption, as pointed out by Hide, is good enough to cater for most cases of physical interest and it merely signifies that the density variation in the fluid is a good deal less than the average density of the fluid. Thus neglecting $a^2$, we can rewrite equation (28) in the form

$$T_1 n^3 + n^2[1 + T_2(k^2 + 1)] + n\left[k^2 + 1 - \frac{GT_1k^2}{1 + k^2}\right] - \frac{Gk^2}{1 + k^2} = 0 \,. \tag{34}$$

A. Unstable Arrangement

If $G > 0$, i.e., if the density of the fluid increases upwards, at least one value of $n$ given by equation (34) is positive, thereby, implying instability. It can be remarked that this conclusion remains unaltered even if we dispense with the restriction (33). Hence we observe that the basic criterion of instability is independent of non-newtonian parameters.

It is evident from (34) that there would be precisely one value of $n$ which is positive. The behaviour of this root, which causes instability of the arrangement, for small and large $k$ is as follows:

$$n \to Gk^2 \,, \qquad k \to 0 \tag{35}$$

and

$$n \to G/k^2 \,, \qquad k \to \infty \,. \tag{36}$$

It follows that a mode of maximum instability exists which is likely to assert itself in the initial stage of the motion.

It can be further seen that the growth rate of disturbance remains essentially stationary for small and large wave-lengths for different values of the non-newtonian parameters $T_1$ and $T_2$. However, one expects these parameters to influence the mode of maximum instability. For this reason, in figure 1, $n$ is plotted against $k$ for a fixed value of $G$ and various values of $T_1$ and $T_2$. We observe that (i) $n$ increases with an increase in the relaxation time $T_1$, (ii) $n$ decreases with an increase in the retardation time $T_2$. The dotted lines indicate the value of $n$ for the newtonian case. It can be seen that this value of $n$ is always less than the corresponding value for non-vanishing values of $T_1$ and $T_2$. One can, therefore, conclude that an unstable non-newtonian arrangement departs at a faster rate from equilibrium than the corresponding newtonian arrangement (i.e., one having the same values of $\varrho_0$, $\eta_0$ and $\beta$).

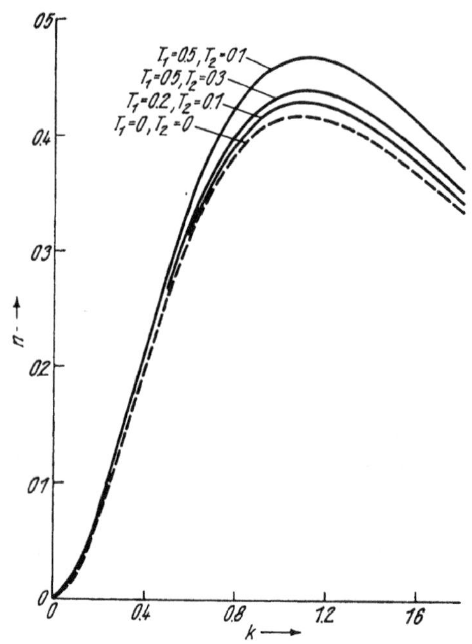


Fig. 1. Illustrating the influence of non-newtonian parameters in the unstable case. The growth rate $n$ is plotted against the wave-number $k$ for $G = 2$ and several values of $T_1$ and $T_2$.

To obtain an expression for $n_m$, the maximum growth rate of disturbance, we differentiate equation (28) with respect to $k$ and set $dn/dk = 0$, which gives

$$T_1 n_m (n_m^2 - G)^2 - 4n_m^2 G T_2 + (n_m^2 - G)^2 - 4n_m G = 0 \,. \tag{37}$$

B. Stable Arrangement

When $G \leqq 0$, all the roots of equation (28) are either real and negative or complex conjugate with negative real parts. In either case the stability of the arrangement is ensured. In the ensuing analysis we shall assume $G = 0$ which means that the buoyancy forces are absent. The reasons for this assumption are twofold (i) the presence of parameter $G$ would obscure the aim of the analysis which is to study the effects of variation of $T_1$ and $T_2$, (ii) it reduces the degree of equation (34), making it a quadratic and thereby simplifies the analysis.

Setting $G = 0$ in equation (28), it simplifies to

$$T_1 n^2 + [1 + T_2(k^2 + 1)]\, n + k^2 + 1 = 0\,. \tag{38}$$

For a newtonian fluid, the above equation further simplifies to

$$n = -(k^2 + 1) \tag{39}$$

which implies that in the absence of buoyancy forces, a newtonian fluid configuration is restored to its equilibrium position in an aperiodic manner, or, in other words, waves can not be excited in this case.

When $T_1 \neq 0$, the solution of equation (38) is

$$n = \frac{1}{2T_1}\left[-1 - T_2(k^2 + 1) \pm \{[1 + T_2(k^2 + 1)]^2 - 4T_1(k^2 + 1)\}^{1/2}\right]. \tag{40}$$

The motion would be oscillatory or aperiodic according as the quantity under the square root sign is negative or positive. To investigate it further we introduce a quantity $z$ defined as follows

$$z = \frac{1}{4}\,\frac{[1 + T_2(k^2 + 1)]^2}{k^2 + 1}\,. \tag{41}$$

Oscillations would take place only for those disturbances for which $T_1 > z$, otherwise the equilibrium would be restored aperiodically. $z$ has a minima or not according as $T_2$ is less than or greater than unity. In fact if $T_2 < 1$ (figure 2), $z$ falls from its initial value $\frac{1}{4}(1 + T_2)^2$ at $k = 0$ to $z = T_2$ at $k = (1/T_2 - 1)^{1/2}$ and then rises to infinity with $k$. On the other hand, if $T_2 > 1$ (figure 3), $z$ rises monotonically. We can now summarise the results as follows

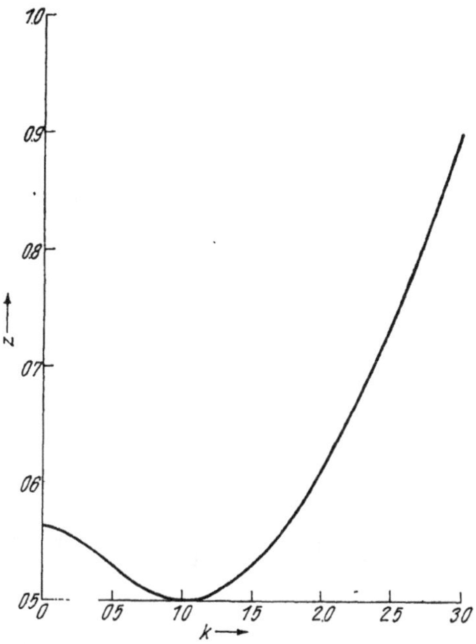


Fig 2. Illustrating the manner in which the equilibrium is restored in the absence of buoyancy forces ($G = 0$). $z$ is plotted as a function of $k$ for $T_2 = 0.5$. Oscillations arise only if $z > T_1$, otherwise the aperiodic motion takes place

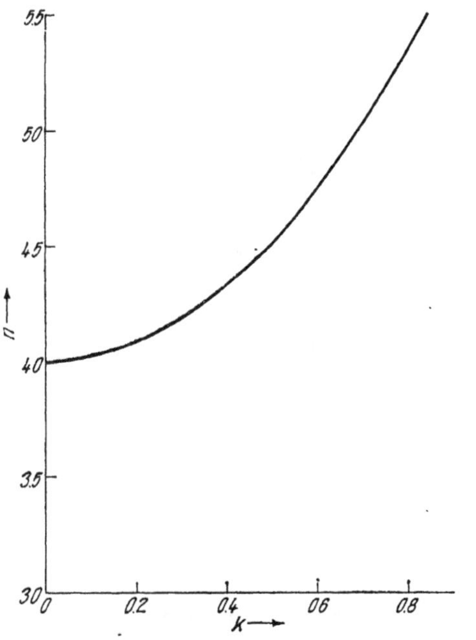


Fig. 3. Illustrating the manner in which the equilibrium is restored in the absence of buoyancy forces ($G = 0$). $z$ is plotted as a function of $k$ for $T_2 = 3.0$. Oscillations arise only if $z > T_1$, otherwis ethe aperiodic motion takes place. (Instead of $n$ read $z$.)

(a) $T_1 > \frac{1}{4}(1 + T_2)^2$. In this case a line drawn parallel to $k$ axis at a height $T_1$ intersects the curve $z$ in one point say $k_2$. Periodic motion is possible only in the range $0 < k < k_2$.

(b) $T_2 < T_1 < \frac{1}{4}(1 + T_2)^2$.

(i) $T_2 < 1$. Now the line cuts the curve in two points say $k_1$ and $k_2$. Oscillations take place in the range $k_1 < k < k_2$, aperiodic motion takes place otherwise.

(ii) $T_2 > 1$. For the values of $T_1$ and $T_2$ lying in this range, the line drawn in figure 3 at a height $T_1$ lies below the curve $z$ for all values of $k$. Hence the equilibrium is restored aperiodically for all disturbances.

The aforementioned results are illustrated in figures 4 to 6 in which $n$ has been plotted against $k$ for different pairs of values of $T_1$ and $T_2$. We conclude this section by remarking that one effect of non-newtonian parameters is to excite waves for some disturbances for which no oscillatory motion is possible in a newtonian fluid.

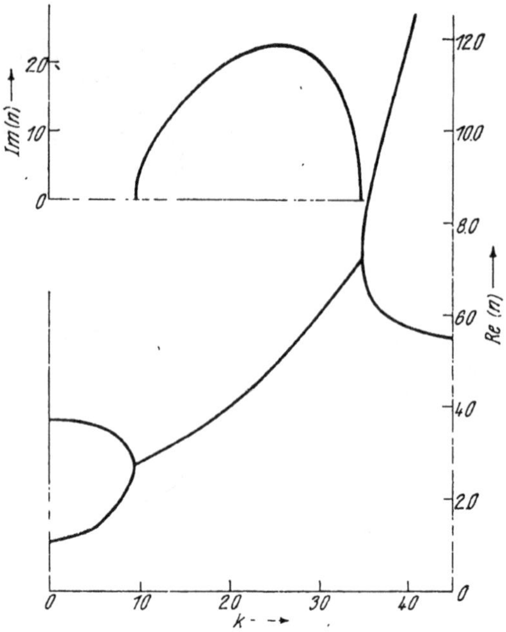


Fig. 4. Illustrating the variation of $n$ with $k$ for $G = 0$, $T_1 = 0.25$ and $T_2 = 0.20$. Waves arise within the range $k_1 < k < k_2$

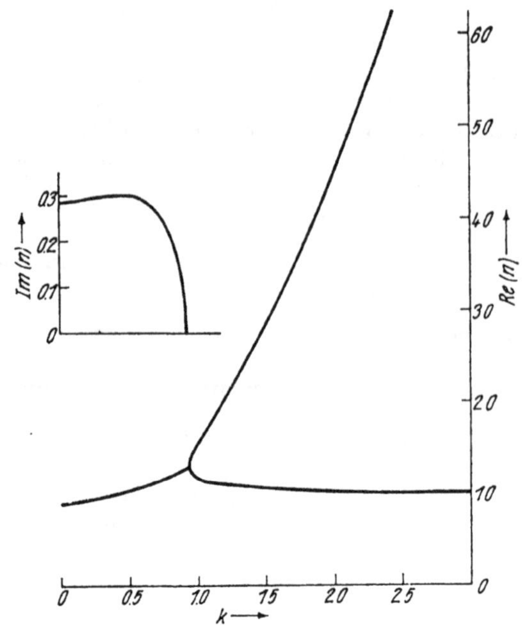


Fig. 5. Illustrating the variation of $n$ with $k$ for $G = 0$, $T_1 = 1.1$ and $T_2 = 1.0$ Waves arise within the range $0 < k < k_2$.

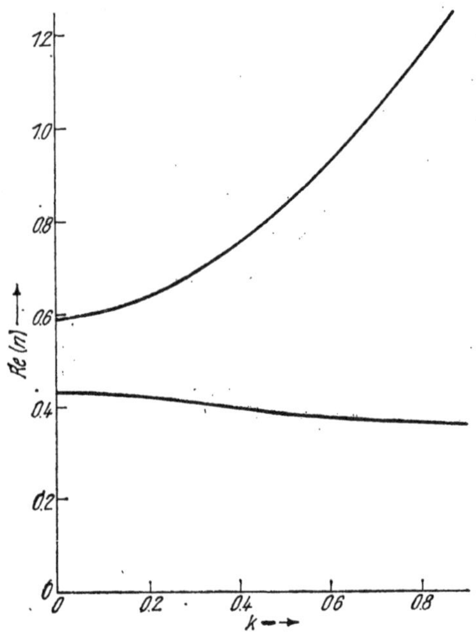


Fig. 6. Illustrating the variation of $n$ with $k$ for $G = 0$, $T_1 = 3.9$ and $T_2 = 3.0$. Aperiodic motion takes place for all disturbances

### Effect of Rotation

In the present section we shall briefly consider the effect of rotation on the stability of Oldroyd fluid. If we assume that the fluid is rotating uniformly about $z$-axis with angular velocity $\Omega$, equations (10) and (11) are modified to

$$n\varrho u - 2\varrho\Omega v = -ik_x\delta p + \mu(D^2 - k^2)\, u + D\mu(Du + ik_x w) \tag{42}$$

$$n\varrho v + 2\varrho\Omega u = -ik_y\delta p + \mu(D^2 - k^2)\, v + D\mu(Dv + ik_y w)\,. \tag{43}$$

We introduce $\zeta$, the $z$-component of vorticity vector

$$\zeta = ik_x v - ik_y u\,. \tag{44}$$

Eliminating some of the perturbed quantities from equations (12), (13), (14), (42) and (43), we arrive at the following pair of differential equations in $w$ and $\zeta$:

$$n[k^2\varrho w - D(\varrho Dw)] - \frac{gk^2}{n}(D\varrho)\, w - D(2\varrho\Omega\zeta) + \mu(D^2 - k^2)^2 w + 2D\mu(D^2 - k^2)\, Dw + {} + D^2\mu(D^2 + k^2)\, w = 0 \tag{45}$$

and

$$[n\varrho - \mu (D^2 - k^2) - D\mu D]\zeta = 2\varrho\Omega Dw \,. \tag{46}$$

The additional boundary conditions on $\zeta$ are

$$\zeta = 0 \qquad \text{on a rigid boundary} \tag{47}$$

and

$$D\zeta = 0 \qquad \text{on a free boundary} \,. \tag{48}$$

To deduce (48) we observe that $ik_x\tau_{yz} - ik_y\tau_{xz} = \mu D\zeta$ and since $\tau_{xz}$ and $\tau_{yz}$ vanish on a free boundary, so does $D\zeta$.

Multiplying equation (45) for subscript $i$ by $w_j$ and integrating across the vertical extent of the fluid, we obtain

$$n_i \int_L \varrho(Dw_i\, Dw_j + k^2 w_i w_j)\, \mathrm{d}z - \frac{gk^2}{n_i} \int_L D\varrho w_i w_j\, \mathrm{d}z +$$
$$+ \int_L \mu[(D^2 + k^2)\, w_i (D^2 + k^2)\, w_j + 4k^2\, Dw_i\, Dw_j]\, \mathrm{d}z + n_j \int_L \varrho\zeta_i\zeta_j\, \mathrm{d}z + \int_L \mu(D\zeta_i\, D\zeta_j + k^2\zeta_i\zeta_j)\, \mathrm{d}z = 0 \,. \tag{49}$$

On setting $i = j$ in equation (49) we obtain a variational formulation of the problem. Indeed on proceeding along the same lines as before it can be shown that the variational principle exists for a rotating configuration as well.

For a fluid having density and viscosity distributions according to equations (23) and (24), the dispersion relation under the restriction $|\beta \mathrm{d}| \ll 1$ becomes (in dimensionless form)

$$T_1^2 n^5 + 2T_1 n^4[1 + T_2(1 + k^2)] + n^3 \left\{[1 + T_2(1 + k^2)^2]^2 + 2T_1(1 + k^2) + T_1^2 \frac{U - Gk^2}{k^2 + 1}\right\} +$$
$$+ n^2 \left\{2(1 + k^2)\,[1 + T_2(1 + k^2)] - GT_1T_2k^2 + 2T_1 \frac{U - Gk^2}{1 + k^2}\right\} +$$
$$+ n \left\{(1 + k^2)^2 - Gk^2(T_1 + T_2) + \frac{U - Gk^2}{1 + k^2}\right\} - Gk^2 = 0 \tag{50}$$

where

$$U = \frac{4\Omega^2 \mathrm{d}^4}{\pi^4 s^4 \nu^2} \tag{51}$$

is a dimensionless measure of CORIOLIS forces in terms of viscous forces.

Clearly if $G > 0$ i.e., if the density of the fluid increases upwards, equation (50) admits one real positive root representing the unstable mode. On the other hand, if $G < 0$, all the terms of equation (50) are positive and hence by HURWITZ' criterion all the roots have negative real parts implying stability. We thus conclude that CORIOLIS forces cannot change the stability character of a potentially unstable stratification.

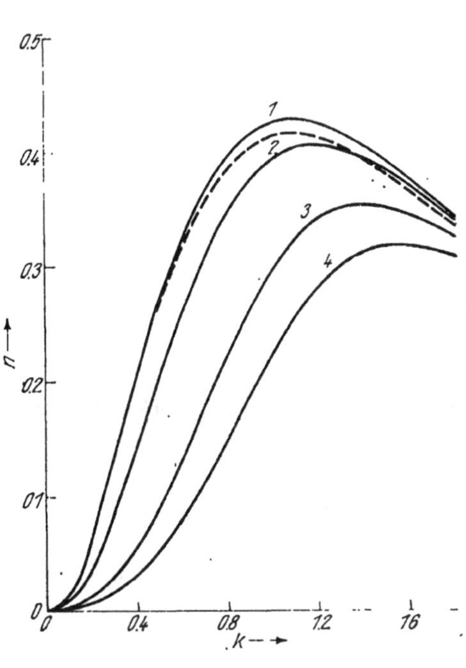


Fig. 7. Illustrating the inhibiting influence of rotation in the unstable case. The growth rate $n$ is plotted against the wave number $k$ for $G = 0$, $T_1 = 0.2$, $T_2 = 0.1$ and several values of $U$. The non-dimensional measure of the effects of rotation. For curves marked 1, 2, 3, 4, $U$ has the values 0, 1, 5 and 10 respectively. The dotted line exhibits the value of $n$ for a newtonian non-rotating fluid

To study the influence of rotation on the growth-rate of disturbance, keeping $G$, $T_1$ and $T_2$ fixed, positive root of $n$ has been plotted against $k$ for different values of $U$ in figure 7. For the sake of comparison, the values of $n$ have also been indicated for newtonian non-rotating configuration by dotted lines. It can be seen from the diagram that (i) $n$ decreases with increasing $U$ for all $k$ (ii) $n_m$ also decreases with increasing $U$ (iii) $k_m$, the wave-number for which the mode of maximum instability occurs increases with increasing $U$. It can also be noted that for non-vanishing $U$, the curves of $n$ are intersecting the corresponding curves for newtonian non rotating fluid. The reason for it lies in the asymptotic behaviour of $n$. We have

$$n \to Gk^2/(1 + U)\,, \qquad k \to 0 \tag{52}$$

$$n \to \frac{G}{k^2}\left\{1 + \frac{1}{k^2}[G(T_1 - T_2) - 2] + \dots\right\}, \qquad k \to \infty\,. \tag{53}$$

For small values of $k$, $n$ is less for rotating configuration than for non-rotating configuration, and for large $k$, irrespective of values of $U$, $n$ is more for newtonian configuration than for non-newtonian configuration.

When $G = 0$, equation (50) becomes a quartic, which in turn splits into two quadratic equations with complex coefficients

$$T_1 n^2 + \left\{1 + T_2(k^2 + 1) \pm i T_1 \sqrt{\frac{U}{k^2 + 1}}\right\} n + k^2 + 1 \pm i \sqrt{\frac{U}{k^2 + 1}} = 0\,. \tag{54}$$

Each of the above equations will give rise to complex roots with negative real parts and non-vanishing imaginary parts. The effect of rotation in this case is, therefore, to excite waves for all disturbances.

### A Note on the Usefulness of Variational Principle-Comparison of Exact and Approximate Results

Though the approximate results obtained by HIDE [4] for the case of two superposed fluids were fairly in agreement with the exact results derived by CHANDRASEKHAR [3], REID [6] indicated that HIDE ommitted an important term and his proximity of approximate and exact results was fortuitous. This naturally raised serious doubts about the usefulness of the variational principle. Here, therefore, we present the results of exact analysis and the analysis due to the variational principle.

Specifically, we are dealing with the case of a fluid in which density varies exponentially with $z$. Substituting for $\varrho$ from equation (23) and for $\mu$ from equations (15) and (24) in equation (16) we obtain

$$[D^4 + 2\beta D^3 + (\beta^2 - b^2)\, D^2 - \beta b^2 D + c]\, w = 0 \tag{55}$$

where

$$b^2 = 2k^2 + \frac{(1 + \lambda_1 n)}{\nu(1 + \lambda_2 n)} \tag{56}$$

and

$$c = k^2\left[\beta^2 + k^2 - \left(\frac{g\beta}{n} - n\right)\frac{1 - \lambda_1 n}{\nu(1 + \lambda_2 n)}\right]. \tag{57}$$

Equation (55) admits a general solution of the form

$$w = \sum_{i=1}^{4} w_i\, e^{q_i z} \tag{58}$$

where $w_i$ are the constants of integration and $q_i$ are the roots of the equation

$$q^4 + 2\beta q^3 + (\beta^2 - b^2)\, q^2 - \beta b^2 q + c = 0\,. \tag{59}$$

Setting

$$q = -\tfrac{1}{2}\beta + m$$

we obtain

$$m^4 - (\tfrac{1}{2}\beta^2 + b^2)\, m^2 + (c + \tfrac{1}{4}\beta^2 b^2 + \tfrac{1}{16}\beta^4) = 0 \tag{60}$$

whose solution is

$$m_{1,2}^2 = \tfrac{1}{4}\beta^2 + \tfrac{1}{2} b^2 \pm (\tfrac{1}{4} b^4 - c)^{1/2}\,. \tag{61}$$

In the present paper we have assumed the bounding planes to be free, where the boundary conditions (17) and (19) must be satisfied. Making use of these boundary conditions for the general solution of $w$ given by equation (58) and eliminating $w_i$, we obtain the following dispersion relation

$$(m_1^2 - m_2^2) \sinh m_1 d \sinh m_2 d - \beta^2[(m_1^2 + m_2^2) \sinh m_1 d \sinh m_2 d + 2 m_1 m_2 (1 - \cosh m_1 d \cosh m_2 d)] = 0\,. \tag{62}$$

Under the assumption $|\beta d| \ll 1$, which has been made in the approximate solution, equation (62) admits a very simple solution

$$md = \pi s \tag{63}$$

where $m$ stands for either $m_1$ or $m_2$.

When one substitutes for $m$ in equation (60), the dispersion relation turns out to be same as equation (27). Hence we see that the approximate results derived in the present paper are in agreement with the exact results.

We now consider the case in which condition (33) is not satisfied. For this purpose, we write

$$m_1 d = x \quad \text{and} \quad m_2 d = iy \tag{64}$$

and assume $x$ and $y$ to be real. We can now rewrite equation (62) in the non-dimensional form as

$$(x^2 - y^2)^2 \sin y \sinh x - \pi^2 a^2[(x^2 - y^2) \sin y \sinh x + 2xy(1 - \cosh x \cos y)] = 0 \tag{65}$$

where $a$ is defined by the equation (32). Here and in the sequel $s$ has been taken equal to unity for the reason that the mode of maximum instability corresponds to $s = 1$.

For any assigned $x$, equation (65) allows infinitely many solutions for $y$. We shall, however, seek only that solution which lies in the interval $(\pi/2, \pi)$ because corresponding to $s = 1$, in the limit as $a \to 0$, the solution for $y$ is $\pi$. In the subsequent analysis we shall measure $n$ and $k$ in terms of $(\pi^2\nu/d^2)$ sec$^{-1}$ and $(\pi/d)$ cm$^{-1}$ so as to make the comparison of exact and approximate solutions easier.

By virtue of definition of $x$ and $y$ given by equation (64), $x^2$ and $-y^2$ are the roots of the equation (see equation (61))

$$\sigma^4 - (\tfrac{1}{2}\pi^2 a^2 + b^2 d^2)\,\sigma^2 + (cd^4 + \tfrac{1}{4}\pi^2 a^2 b^2 d^2 + \tfrac{1}{16}\pi^4 a^4) = 0\,. \tag{66}$$

This leads to the following

$$\xi \equiv x^2 - y^2 = \pi^2 \left[\frac{1}{2} a^2 + 2k^2 + \frac{(1 + T_1 n)}{1 + T_2 n}\right] \tag{67}$$

$$\eta^2 \equiv 4x^2y^2 = -\pi^4 \left[4k^4 + 2k^2 \left\{2\left(n - \frac{G}{n}\right)\frac{1 + T_1 n}{1 + T_2 n} + 3a^2\right\} + \frac{1}{4} a^4 + \frac{a^2 n(1 + T_1 n)}{1 + T_2 n}\right]. \tag{68}$$

We shall limit ourselves to the discussion of unstable arrangement. Thus only the positive values of $n$ will be admissible. From equation (67) we have

$$k^2 = \frac{1}{2}\left[\frac{\xi}{\pi^2} - \frac{1}{2} a^2 - \frac{n(1 + T_1 n)}{1 + T_2 n}\right]. \tag{69}$$

On substituting the value of $k^2$ from equation (69) in equation (68), we obtain the following fifth degree equation in $n$:

$$\begin{aligned} &T_1^2 n^5 + 2T_1 n^4 (1 + a^2 T_2) + n^3 \left\{1 + 2a^2(T_1 + T_2^2) - 2GT_1 - \left(\frac{a^2\xi}{\pi^2} - a^4 + \frac{\xi^2 + \eta^2}{\pi^4}\right) T_2^2\right\} + \\ &+ n^2 \left\{2a^2 - 4GT_1 - 2T_2\left(\frac{2a^2\xi}{\pi^2} - a^4 + \frac{\xi^2 + \eta^2}{\pi^4}\right) + GT_1T_2\left(\frac{2\xi}{\pi^2} - a^2\right)\right\} + \\ &+ n \left\{G(T_1 + T_2)\left(\frac{2\xi}{\pi^2} - a^2\right) - 2G - \left(\frac{2a^2\xi}{\pi^2} - a^4 + \frac{\xi^2 + \eta^2}{\pi^4}\right)\right\} + G\left(\frac{2\xi}{\pi^2} - a^2\right) = 0\,. \end{aligned} \tag{70}$$

To obtain the complete $(k, n)$ relationship, we proceed as follows: We assign a value to $x$, obtain from equation (62) the value of $y$ which lies in the interval $(\pi/2, \pi)$. Next $\xi$ and $\eta$ are calculated from (67) and (68) by substituting for $x$ and $y$. Now the five roots of equation (70) are computed, out of which the smallest positive root is chosen as this will ensure that the right hand side of equation (69) does not become negative. Finally $k$ is calculated from equation (69). The results of these calculations, which were carried out on CDC 6000 at the University of Calgary, are presented in Table 1, in which the values of $n$ computed from equation (28) are also given. It can be seen that these results are in fair agreement, which justifies the use of variational principle.

To get further refined value of $n$ by using variational principle, we assume

$$w = W(\sin lz + \alpha \sin 2lz) \tag{71}$$

where $\alpha$ is a parameter. The dispersion relation is obtained by substituting for $w$ in variational formulation (21). Since $n$ is stationary with respect to small variations $\delta w$ in $w$, we have obtained the stationary value of $n$ on varying $\alpha$. This value is also given in Table 1 and a comparison with the exact value shows that this value is very close to the exact value.

The foregoing discussion should dispel doubts about using variational principle for investigating the Rayleigh Taylor instability of a fluid.

### Acknowledgement

Financial support of the National Research Council of Canada is gratefully acknowledged.

Table No. 1: Illustrating the variation of $n$, the growth rate of disturbance with $k$, wave-number of disturbance for $G = 2.0$, $T_1 = 0.2$, $T_2 = 0.1$ and various values of $a$

| $a = 0.1$ | | | | $a = 0.5$ | | | |
|---|---|---|---|---|---|---|---|
| $k$ | exact $n$ | approx. $n$ | improved $n$ | $k$ | exact $n$ | approx. $n$ | improved $n$ |
| 0.0955 | .0177 | .0176 | .0176 | .1813 | .0601 | .0574 | .0575 |
| .2131 | .0780 | .0777 | .0777 | .2692 | .1155 | .1107 | .1109 |
| .2963 | .1337 | .1334 | .1335 | .3446 | .1657 | .1594. | .1599 |
| .4377 | .2302 | .2297 | .2299 | .4810 | .2497 | .2412 | .2420 |
| .5674 | .3058 | .3053 | .3055 | .6109 | .3124 | .3025 | .3040 |
| .6920 | .3609 | .3604 | .3606 | .7356 | .3556 | .3451 | .3471 |
| .8135 | .3974 | .3969 | .3972 | .8571 | .3824 | .3717 | .3742 |
| .9323 | .4185 | .4178 | .4182 | .9755 | .3963 | .3857 | .3889 |
| 1.0486 | .4272 | .4266 | .4270 | 1.0911 | .4002 | .3900 | .3943 |
| 1.1623 | .4266 | .4260 | .4264 | 1.2041 | .3970 | .3875 | .3914 |
| 1.2737 | .4195 | .4189 | .4193 | 1.3147 | .3887 | .3799 | .3834 |
| 1.3828 | .4078 | .4074 | .4077 | 1.4231 | .3771 | .3689 | .3720 |
| 1.4899 | .3933 | .3928 | .3931 | 1.5295 | .3634 | .3559 | .3584 |
| 1.5952 | .3770 | .3766 | .3769 | 1.6342 | .3485 | .3416 | .3438 |
| 1.6989 | .3599 | .3596 | .3599 | 1.7374 | .3331 | .3268 | .3288 |

## References

1 RAYLEIGH, Lord, Investigation of the character of the equilibrium of an incompressible heavy fluid of variable density, Proc. Lond. Math. Soc., 14, 170 (1883), also Scientific papers 2, 200 (1900).
2 HARRISON, W. T., The influence of viscosity of the oscillations of superposed fluids, Proc. Lond. Math. Soc. 6, 396 (1908).
3 CHANDRASEKHAR, S., Hydrodynamic and Hydromagnetic Stability, Oxford University Press, London (1961).
4 HIDE, R., The character of equilibrium of an incompressible heavy viscous fluid of variable density: an approximate theory, Proc. Camb. Phil. Soc. 51, 179 (1955).
5 OLDROYD, J. G., Non-Newtonian effects in steady motion of some idealized elasto-viscous fluids, Proc. Roy. Soc. (London) A 245, 278 (1958).
6 REID, W. J., The effects of surface tension and viscosity on the stability of two superposed fluids, Proc. Camb. Phil. Soc., 57, 415 (1961).

Eingereicht am 5. 6. 1978

*Anschrift*: P. D. ARIEL (Permanent Address: Department of Mathematics, Malaviya Regional Engineering College, Jaipur, India) and Prof. B. D. AGGARWALA, Department of Mathematics and Statistics, Faculty of Science, The University of Calgary, 2920 24 Ave. N.W., Calgary, Canada T2N 1N4

# KLEINE MITTEILUNGEN

ZAMM 60, 49 – 50 (1980)

U. FISCHER

## Bestimmung der kleinsten Eigenwerte symmetrischer, positiv definiter Matrizen durch direkte Vektoriteration

### 0. Problemstellung

Zur Bestimmung der kleinsten Eigenfrequenzen elastischer Bauteile und der zugehörigen Eigenschwingungsformen mit Hilfe der Methode der finiten Elemente ist die Lösung des Matrizen-Eigenwertproblems für große symmetrische Bandmatrizen notwendig. In der Regel sind nur die kleinsten Eigenfrequenzen von technischer Bedeutung; die höchsten Eigenfrequenzen des Finite-Element-Modells hängen ohnehin nur vom Diskretisierungsverfahren ab und sind damit nicht Eigenschaften des dem Modell zur Grunde liegenden Kontinuums. Als ein geeignetes und effektives Verfahren zur Bestimmung eines schmalen Bandes aus dem Spektrum der Eigenwerte bei großen Bandmatrizen hat sich die VON MISESsche Vektoriteration erwiesen. Diese kann durch eine Unterraum-Transformation in der Konvergenz noch beschleunigt werden, vgl. [1]. Ein Nachteil ist jedoch, daß die Vektoriteration bei direkter Anwendung, dargestellt durch

$$z^{(i+1)} = Cz^{(i)} \tag{1}$$

gegen den Eigenvektor $z$ mit dem höchsten Eigenwert der Matrix $C$ konvergiert. Dagegen konvergiert die sogenannte inverse Vektoriteration, dargestellt durch die Gl.

$$Cz^{(i+1)} = z^{(i)} \tag{2}$$

gegen den Eigenvektor mit dem kleinsten Eigenwert. Hierbei wurde der Einfachheit halber vorausgesetzt, daß alle Eigenwerte voneinander verschieden sind. Sie seien so geordnet, daß

$$\lambda_1 < \lambda_2 < \dots < \lambda_{n-1} < \lambda_n \tag{3}$$

gilt. Bei großen symmetrischen Bandmatrizen geschieht die Auflösung der Gl. (2) nach $z^{(i+1)}$ am effektivsten durch die einmalige CHOLESKY-Zerlegung der Matrix und anschließendes sog. „Vorwärtseinsetzen" und „Rückwärtseinsetzen", vgl. z. B. [2]. Bei diesem Vorgehen kann die Bandstruktur der Matrix $C$ genutzt werden. Nachteilig ist neben dem Aufwand für die CHOLESKY-Zerlegung vor allem der Umstand, daß bei diesem Verfahren die relativ dünne Besetzung der Matrix mit von Null verschiedenen Elementen auch innerhalb des Bandes nicht genutzt werden kann. Das im folgenden vorgestellte Verfahren vermeidet diesen Nachteil.

### 1. Direkte Vektoriteration mit dominantem ersten Eigenvektor

Es sei $x$ ein Eigenvektor der Matrix $C$ mit dem Eigenwert $\lambda$. Weiterhin sei eine Iterationsvorschrift gegeben durch

$$z^{(i+1)} = \bar{\lambda}_{i+1} z^{(i)} - Cz^{(i)} . \tag{4}$$

Wendet man diese $m$-mal auf den „Startvektor" $z^{(0)} = x$ an, so erhält man

$$z^{(m)} = \prod_{i=1}^{m} (\bar{\lambda}_i E - C)\, z^{(0)} = \prod_{i=1}^{m} (\bar{\lambda}_i - \lambda)\, x . \tag{5}$$

Es kommt darauf an, die $\bar{\lambda}_i$ so zu wählen, daß das Produkt

$$P_m(\lambda) = \prod_{i=1}^{m} (\bar{\lambda}_i - \lambda) \tag{6}$$

folgende Eigenschaften hat:

$$P_m(\lambda_1) > P_m(\lambda_2) > \dots > P_m(\lambda_l) > \sup_{k>l} |P_m(\lambda_k)| . \tag{7}$$

In diesem Falle dominiert bei genügend großem $m$ nach $m$-facher Iteration nach Gl. (4) der erste Eigenwert, wenn man einen Startvektor $z^{(0)}$ wählt, der nicht orthogonal zu diesem ist. Wählt man dagegen einen zum ersten Eigenvektor orthogonalen Startvektor, so dominiert der zweite Eigenvektor usw.

Man erreicht die geforderte Eigenschaft (7) des Polynoms (6), wenn man seine Wurzeln $\bar{\lambda}_i$ auf die Nullstellen der ČEBYŠEVschen Polynome

$$T_m = \cos (m \arccos t) , \qquad t \in [-1, 1]$$

linear abbildet, so daß

$$\bar{\lambda}_i = \frac{\lambda'' - \lambda'}{2} \cdot \frac{\cos \frac{2i-1}{2m} \pi}{\cos \frac{\pi}{2m}} + \frac{\lambda'' + \lambda'}{2} \tag{8}$$

ist. Für die Konstanten $\lambda''$ und $\lambda'$ gilt

$$\lambda'' \geqq \lambda_n , \tag{9}$$

$$\lambda' > \lambda_l . \tag{10}$$

### 2. Eine zweistufige Rekursionsformel

Die Iteration nach Gl. (4) mit den $\bar{\lambda}$-Werten nach Gl. (8) hat den Nachteil, daß die Anzahl der Iterationen $m$ von vornherein festgelegt werden muß, ohne daß man schon beurteilen kann, ob sie ausreichend sein wird. Dazu kommen numerische Schwierigkeiten durch Rundungsfehler und mögliche Zahlenwertüberschreitungen im Rechner während der Abarbeitung, u. a. deshalb, weil nach Ausführung eines Teiles der Iterationen noch Eigenvektoren der höheren Eigenwerte dominant sein können. Diesen Nachteil vermeidet die Anwendung der Rekursionsformel

$$T_{m+1}(t) + T_{m-1}(t) = 2t \cdot T_m(t) , \qquad T_{-1} = T_1 , \tag{11}$$

der ČEBYŠEVschen Polynome. Es sei $x$ wieder ein Eigenvektor der Matrix $C$ mit dem Eigenwert $\lambda$. Weiter sei

$$x_m = \alpha^m T_m x \tag{12}$$

definiert. Dabei ist $\alpha$ ein Normierungsfaktor, über den frei verfügt werden kann. Bei Anwendung der Gleichung (11) erhält man

$$x_{m+1} = 2\alpha^{-1} t x_m - \alpha^{-2} x_{m-1} , \qquad x_{-1} = x_1 , \qquad x_0 = x .$$

Zur Erzielung der gewünschten Dominanzeigenschaften der Polynome $T_m$ bzw. der ihnen proportionalen $P_m$ ist $t$ in Abhängigkeit von $\lambda$ so zu bestimmen, daß $\lambda'$ auf $t = 1$ und $\lambda''$ auf $t = -1$ abgebildet wird. Das ist durch

$$t x_m = \left( \frac{\lambda'' + \lambda'}{\lambda'' - \lambda'} E - \frac{2}{\lambda'' - \lambda'} C \right) x_m \tag{13}$$

zu erreichen. Für beliebige Startvektoren erhält man so die gewünschte Rekursionsformel

$$z_{m+1} = 2\alpha^{-1} \left( \frac{\lambda'' + \lambda'}{\lambda'' - \lambda'} E - \frac{2}{\lambda'' - \lambda'} C \right) z_m - \alpha^{-2} z_{m-1}$$
$$z_{-1} = z_{+1} \tag{14}$$

### 3. Konvergenzverhalten

Die Eigenfrequenzen elastischer Kontinua sind gewöhnlich eng benachbart, deshalb soll als Maß für die Konvergenzgüte der Wert

$$\eta = \left| \frac{1}{T_m} \cdot \frac{\mathrm{d}T_m}{\mathrm{d}\lambda} \right| = \left| \frac{\mathrm{d} \ln T_m}{\mathrm{d}\lambda} \right| \tag{15}$$

angesehen werden. Mit

$$t = 1 + 2 \frac{\lambda' - \lambda}{\lambda'' - \lambda'} > 1 \tag{16}$$

für $\lambda < \lambda' \ll \lambda''$ (vgl. Gl. (13)) nehmen die ČEBYŠEVschen Polynome die Form

$$T_m = \cosh (m \operatorname{arcosh} t) \tag{17}$$

an, woraus schließlich

$$\eta = m \frac{\tanh \left[ m \operatorname{arcosh} \left( 1 + 2 \frac{\lambda' - \lambda}{\lambda'' - \lambda'} \right) \right]}{\sqrt{(\lambda'' - \lambda)(\lambda' - \lambda)}} \tag{18}$$

folgt. Ist $t$ nach Gl. (16) nur wenig größer als 1, so gilt die Abschätzung

$$\eta \approx m \frac{\tanh 2m \sqrt{\frac{\lambda' - \lambda}{\lambda'' - \lambda'}}}{\sqrt{(\lambda'' - \lambda)(\lambda' - \lambda)}} . \tag{19}$$

### 4. Kombiniertes Verfahren

Es zeigt sich, daß für Verhältnisse $\lambda_n/\lambda_1 > 10^4$, wie sie bei Eigenfrequenzberechnungen von Kontinua nach der Finite-Element-Methode auftreten, keine Beträge von $\eta$ nach Gl. (19) erreicht werden, die mit denen aus der inversen Vektoriteration

$$\eta = \frac{m}{\lambda} \tag{20}$$

konkurrenzfähig wäre. Einen Ausweg bietet die schon erwähnte Subspace-Transformation. Dazu stellt man mit Hilfe einer geeigneten Matrix-Norm $\lambda''$ als obere Schranke von $\lambda_n$ fest und wählt $\lambda'$ als obere Schranke für die zu berechnenden Eigenwerte $\lambda_1$ bis $\lambda_l$. Die Rekursionsformel (14) ist auf hinreichend viele (mindestens $l$) linear unabhängige Startvektoren $\boldsymbol{z}$ anzuwenden. Die Anzahl der Iterationen $m$ kann aus dem Faktor $\varkappa$ bestimmt werden, um den der zu $\lambda_1$ gehörige Eigenvektor gegenüber den Eigenvektoren mit Eigenwerten $\lambda > \lambda'$ hervorgehoben werden soll. Mit Hilfe der Gln. (16) und (17) kann man abschätzen:

$$m \sqrt{\frac{\lambda' - \lambda_1}{\lambda'' - \lambda'}} = \frac{1}{2} \operatorname{arcosh} \varkappa \,. \tag{21}$$

Einige Zahlenwerte ergeben sich aus der folgenden Tabelle:

| $\varkappa$ | 10 | 30 | 100 | 300 | 1000 | 3000 | 10000 |
|---|---|---|---|---|---|---|---|
| $m \sqrt{\frac{\lambda' - \lambda_1}{\lambda'' - \lambda'}}$ | 1,50 | 2,05 | 2,65 | 3,20 | 3,80 | 4,35 | 4,95 |

Nach einigen Iterationen sind die iterierten Spaltenvektoren jeweils zu orthonormieren, um zu verhindern, daß durch die Dominanz der zu den kleinsten Eigenvektoren gehörigen Eigenvektoren und die Rundungsfehler die lineare Unabhängigkeit verloren geht. Im Ergebnis von $m$ Iterationen und anschließender Orthonormierung erhält man eine Rechteckmatrix $\boldsymbol{Z}$, für die

$$\boldsymbol{Z}^T\boldsymbol{Z} = \boldsymbol{E}$$

gilt. Nun ist die Matrix $\boldsymbol{A} = \boldsymbol{Z}^T\boldsymbol{C}\boldsymbol{Z}$ zu bilden. Ihre Eigenwerte stimmen mit den ersten Eigenwerten der Matrix $\boldsymbol{C}$ näherungsweise überein, wobei die Güte der Näherung von $\varkappa$ bestimmt wird. Wegen des erheblich geringeren Formates der Matrix $\boldsymbol{A}$ ist die Bestimmung aller ihrer Eigenwerte und Eigenvektoren verhältnismäßig einfach, z. B. mit dem Jacobi-Verfahren. Sind die Eigenvektoren von $\boldsymbol{A}$ in der Matrix $\boldsymbol{Y}$ zusammengefaßt, so bestimmt man die $l$ Eigenvektoren von $\boldsymbol{C}$ als Spaltenvektoren von

$$\boldsymbol{X} = \boldsymbol{Z} \cdot \boldsymbol{Y} .$$

### 5. Effektivität und Datenorganisation

Auch nach dem unter 4. vorgeschlagenen kombinierten Verfahren wurden im allgemeinen mehr Iterationen benötigt als nach dem Verfahren der inversen Vektoriteration. Dennoch ist zu erwarten, daß es für die Bestimmung der kleinsten Eigenwerte und der zugehörigen Eigenvektoren für große und sehr große Bandmatrizen effektiver arbeitet. Dafür gibt es vor allem folgende Gründe:

1. Die Cholesky-Zerlegung der Matrix $\boldsymbol{C}$, deren Aufwand quadratisch mit der Bandweite steigt, entfällt.
2. Bei Matrizen $\boldsymbol{C}$, die zur Berechnung von Eigenschwingungen von Kontinua dienen, besteht die überwiegende Zahl der Elemente zwischen dem Hauptdiagonalelement einer Zeile und dem äußersten von Null verschiedenen Element aus Nullelementen. Bei dem vorgestellten Verfahren der direkten Iteration kann auf die Multiplikation dieser Elemente verzichtet werden. Dagegen überträgt sich diese Eigenschaft der Matrix $\boldsymbol{C}$ nicht auf die im Ergebnis der Cholesky-Zerlegung erhaltenen Dreieckmatrizen. Bei einer Datenorganisation, die das Submatrizenverfahren [3] verwirklicht, kann die dünne Besetzung von $\boldsymbol{C}$ nur bedingt genutzt werden, und zwar dadurch, daß die Multiplikation mit Nullsubmatrizen nicht elementweise ausgeführt wird. Effektiver arbeitet eine Kompaktdatenverarbeitung, bei der von der Matrix $\boldsymbol{C}$ nur noch die Nichtnullelemente gespeichert werden. In diesem Falle müssen allerdings ihre Indizes in einer gesonderten Matrix erfaßt und gespeichert werden.
3. Durch den Verzicht auf Speicherung von Nullelementen bzw. Nullsubmatrizen entsteht eine Verminderung des Speicherbedarfs, der andererseits zu einer erheblichen Verminderung der für Externspeicheroperationen benötigten Rechenzeit führen kann.

### 6. Literatur


1 Bathe, K.-J.; Wilson, E.-L., Solution methods for eigenvalue problems in structural mechanics, Intern. Journal of Numerical Methods in Engineering, Vol. 6, p. 213–226, 1973.
2 Dankert, J., Numerische Methoden der Mechanik, VEB Fachbuchverlag, Leipzig 1977.
3 Datenverwaltungssystem FEDAM 1977, Technische Hochschule Magdeburg 1977.


Eingereicht am 17. 5. 1979

*Anschrift*: Prof. Dr. Udo Fischer, Technische Hochschule „Otto von Guericke“, Magdeburg, Boleslaw-Bierut-Platz, DDR

ZAMM 60, 50 – 53 (1980)

F. Strehmel

## Bemerkungen zum Verfahren von Nickel und Rieder

### 1. Einleitung

Zur numerischen Berechnung einer Näherungslösung des skalaren Anfangswertproblems

$$y' = f(x, y) \; ; \qquad y(x_0) = y_0 \tag{1.1}$$

in einem kompakten Intervall $I := [x_0, x_N]$ der reellen Zahlengeraden approximieren Nickel und Rieder [1] die Funktion $f(x, y)$ in jedem Integrationsintervall $I_n := [x_n, x_{n+1}]$, $x_{n+1} = x_n + h$; $n = 0, 1, \dots, N - 1$, durch

$$f(x, y) \approx a \cdot y + b + c \cdot (x - x_n) + d \cdot (x - x_n)^2 \tag{1.2}$$

mit geeigneten rellen Konstanten $a, b, c$ und $d$ (Methode der Exponentialanpassung). In einfacher Weise ergeben sich damit für (1.1) zwei neue $L$-stabile Einschrittverfahren, die als Verallgemeinerungen des klassischen Halbschrittverfahrens (Verfahren von Heun) und des klassischen Runge-Kutta-Verfahrens vierter Ordnung angesehen werden können. Ein Beweis für die Genauigkeitsordnung des „Neuen Runge-Kutta-ähnlichen Verfahrens“ wird nicht gegeben.

Die vorliegende Arbeit enthält erstens eine *allgemeine Konsistenzuntersuchung für Verfahren, die auf einer Approximation von $f(x, y)$ in der Form* (1.2) *beruhen*. Damit ergeben sich eine Klasse von Einschrittverfahren mit unterschiedlicher Konsistenzordnung. Insbesondere erhält man ein Verfahren 5. Ordnung, für das in jedem Integrationsintervall $I_n$, $n = 0, 1, \dots, N - 1$, eine Berechnung der partiellen Ableitung $f_y(x, y)$ und vier Berechnungen von $f(x, y)$ erforderlich sind. Das ist insofern bemerkenswert, da es kein klassisches explizites fünf-stufiges Runge-Kutta-Verfahren der Konsistenzordnung 5 gibt (vgl. Grigorieff [2]). Zweitens wird gezeigt, daß *die Verfahren intern $S$-stabil* (vgl. Verwer [5]) *sind*, so daß sie insbesondere auch zur numerischen Integration steifer Anfangswertprobleme geeignet sind.

### 2. Eine Klasse exponentiell angepaßter Einschrittverfahren

Mit den von Nickel und Rieder [1] eingeführten Hilfsfunktionen

$$e_0(x) = \exp(x); \quad e_{\mu+1}(x) = (\mu e_\mu(x) - 1)/x; \quad \mu = 0, 1, \dots, \tag{2.0}$$

besitzt die Lösung $Y(x)$ des Anfangswertproblems

$$Y'(x) = a \cdot Y(x) + b \,, \quad Y(x_n) = Y_n; \quad n = 0, 1, \dots, N - 1 \,, \tag{2.1}$$

wobei

$$a = f_y(x_n, Y_n) \,, \quad b = f(x_n, Y_n) - a \cdot Y_n; \quad Y_0 = y_0 \,,$$

ist, an der Stelle $x_n + t_1 h$ die Darstellung

$$\overline{Y}_1 := Y(x_n + t_1 h) = e_0(a t_1 h) \, Y_n + b t_1 h e_1(a t_1 h); \quad 0 < t_1 \leqq 1 \,. \tag{2.2}$$

Wegen

$$e_\mu(x) = (\mu - 1)\,! \sum_{i=0}^{\infty} \frac{x^i}{(i + \mu)!}\,; \quad \mu = 1, 2, \dots ,$$

ergibt sich dann für (2.2) die asymptotische Entwicklung

$$\overline{Y}_1 = Y + t_1 fh + t_1^2 af \frac{h^2}{2} + t_1^3 a^2 f \frac{h^3}{6} + O(h^4)\,, \tag{2.3}$$

wobei der Index $n$ zur Kennzeichnung der Ausgangsstelle fortgelassen wurde. Durch Taylor-Entwicklung an der Stelle $(x_n, Y_n)$ erhält man mit (2.3)

$$f(x_n + t_1h, \overline{Y}_1) =$$
$$= f + t_1 Df \cdot h + t_1^2 [D^2f + af_x] \frac{h^2}{2} + t_1^3 [D^3f + a(a^2f - D^2f) -$$
$$- 3f_x Df_y] \frac{h^3}{6} + O(h^4)\,.$$

Dabei bezeichnet $D$ den linearen Differentialoperator

$$D = \frac{\partial}{\partial x} + f \frac{\partial}{\partial y}\,,$$

und es gilt

$$D^0 f = f; \quad D^k f = D(D^{k-1} f); \quad k = 1, 2, \dots, .$$

Aus der Forderung

$$f(x_n + t_1h, \overline{Y}_1) = a \cdot \overline{Y}_1 + b + c^* \cdot t_1 h$$

folgt mit (2.2)

$$c^* h = \frac{1}{t_1} [f(x_n + t_1h, \overline{Y}_1) - f(x_n, Y_n)\, e_0(a \cdot t_1h)]\,.$$

Die Lösung des Anfangswertproblems

$$Y'(x) = a \cdot Y(x) + b + c^* \cdot (x - x_n)\,, \quad Y(x_n) = Y_n\,,$$

an den Stellen $x_n + t_i h$ mit $0 < t_i \leqq 1$ für $i = 2, 3$, lautet:

$$\overline{Y}_i := Y(x_n + t_ih) = e_0(a \cdot t_ih)\, Y_n + bt_ihe_1(a \cdot t_ih) +$$
$$+ c^*(t_ih)^2\, e_2(a \cdot t_ih). \tag{2.4}$$

Daraus erhält man

$$\overline{Y}_i = Y + t_i fh + t_i^2 Df \frac{h^2}{2} + t_i^3 \left[\left(1 - \frac{3}{2}\tau_i\right) aDf + \frac{3}{2}\tau_i D^2 f\right] \frac{h^3}{6} +$$
$$+ t_i^4 [a^2 Df + 2\tau_i a(D^2f - aDf) +$$
$$+ 2\tau_i^2 (D^3 f - aD^2 f - 3f_x Df_y)] \frac{h^4}{24} + O(h^5) \tag{2.5}$$

mit $\tau_i = \frac{t_1}{t_i}$. Unter Benutzung von (2.5) folgt für $f(x_n + t_ih, \overline{Y}_i)$ die asymptotische Entwicklung

$$f(x_n + t_ih, \overline{Y}_i) = f + t_i Df \cdot h + t_i^2 D^2 f \frac{h^2}{2} + t_i^3 \times$$
$$\times \left[D^3 f - \left(1 - \frac{3}{2}\tau_i\right)(D^2f - aDf)\, a\right] \frac{h^3}{6} +$$
$$+ t_i^4 \left[D^4 f - 4\left(1 - \frac{3}{2}\tau_i\right)(D^2f - aDf)\, Df_y -\right.$$
$$\left. - a\left(D^3 f - \frac{d^4}{dh^4}\, Y_i\right)\right] \frac{h^4}{24} + O(h^5)\,.$$

Aus dem linearen Gleichungssystem

$$f(x_n + t_ih, \overline{Y}_i) = a \cdot \overline{Y}_i + b + ct_ih + d(t_ih)^2; \quad i = 2, 3, ,$$

ergeben sich dann mit $t_2 \neq t_3$ in $I_n$ für $c$ und $d$ die Beziehungen

$$c \cdot h = \frac{1}{t_2 t_3 (t_3 - t_2)} [t_3^2 (f(x_n + t_2h, \overline{Y}_2) - a \cdot \overline{Y}_2 - b) -$$
$$- t_2^2 (f(x_n + t_3h, \overline{Y}_3) - a \cdot \overline{Y}_3 - b)]\,, \tag{2.6}$$
$$d \cdot h^2 = \frac{1}{t_2 t_3 (t_3 - t_2)} [t_2 (f(x_n + t_3h, \overline{Y}_3) - a \cdot \overline{Y}_3 - b) -$$
$$- t_3 (f(x_n + t_2h, \overline{Y}_2) - a \cdot \overline{Y}_2 - b)]\,.$$

Daraus folgt

$$c \cdot h = [Df - af]\, h - v_0 [D^3 f - aD^2 f] \frac{h^3}{6} -$$
$$- v_0 v_1 \left[D^4 f - 4\left(1 - \frac{3}{2}\frac{t_1}{v_1}\right)(D^2f - aDf)\, Df_y - aD^3 f\right] \frac{h^4}{24} + O(h^5), \tag{2.7}$$
$$d \cdot h^2 = [D^2 f - aDf] \frac{h^2}{2} + v_1 [D^3 f - aD^2 f] \frac{h^3}{6} +$$
$$+ v_2 \left[D^4 f - aD^3 f - 4\left(1 - \frac{3}{2} t_1 \frac{v_1}{v_2}\right)(D^2 f - aDf)\, Df_y\right] \frac{h^4}{24} + O(h^5),$$

wobei zur Abkürzung

$$v_0 = t_2 t_3; \quad v_1 = t_2 + t_3; \quad v_2 = t_2^2 + t_2 t_3 + t_3^2 \tag{2.8}$$

gesetzt wurde. Definiert man $c$ und $d$ durch die Beziehungen (2.6), so liefert das Anfangswertproblem

$$Y'(x) = a\,Y(x) + b + c \cdot (x - x_n) + d \cdot (x - x_n)^2\,, \quad Y(x_n) = Y_n\,,$$

an der Stelle $x_n + h$ die Lösung

$$\overline{Y}_4 := Y(x_n + h) = e_0(ah)\, Y_n + bhe_1(ah) + ch^2 e_2(ah) + dh^3 e_3(ah)\,. \tag{2.9}$$

Mit (2.7) ergibt sich aus (2.9) für $\overline{Y}_4$ die asymptotische Entwicklung

$$\overline{Y}_4 = Y + fh + Df \frac{h^2}{2} + D^2 f \frac{h^3}{6} + \left[\left(\frac{4}{3} v_1 - 2v_0\right) \times\right.$$
$$\left. \times (D^3 f - aD^2 f) + aD^2 f\right] \frac{h^4}{24} +$$
$$+ \left[5\left(\frac{1}{3} v_2 - \frac{1}{2} v_0 v_1\right)(D^4 f - aD^3 f) + \left(\frac{5}{3}(v_1 - 2v_0) \times\right.\right.$$
$$\left. \times (D^3 f - aD^2 f) + aD^2 f\right) a +$$
$$+ 4\left(\frac{5}{3} v_2 \left(\frac{3}{2}\frac{t_1 v_1}{v_2} - 1\right) - \frac{5}{2} v_0 v_1 \left(\frac{3}{2}\frac{t_1}{v_1} - 1\right)\right) \times$$
$$\left. \times Df_y (D^2 f - aDf)\right] \frac{h^5}{120} + O(h^6)\,. \tag{2.10}$$

Die Lösung $y(x)$ von (1.1) hat an Stelle $x_{n+1} = x_n + h$ die Taylor-Entwicklung

$$y_1 := y(x_n + h) = y(x_n) + \sum_{i=0}^{4} D^i f(x_n, y_n) \frac{h^{i+1}}{(i+1)!} + O(h^6)\,. \tag{2.11}$$

Damit ergeben sich jetzt die folgenden exponentiell angepaßten **Einschrittverfahren**:

1. Setzt man $t_1 = 1$, so liefert bereits (2.2) nach (2.3) und (2.11) ein Verfahren der Konsistenzordnung $p = 1$ (Verallgemeinertes Eulersches Polynomzugverfahren). Pro Integrationsschritt sind zwei Funktionsberechnungen ($f(x, y)$ und $f_y(x, y)$) erforderlich.

2. Mit $t_2 = 1$ und $0 < t_1 \leqq 1$ besitzt die Näherungslösung $\overline{Y}_2$ nach (2.5) und (2.11) die Konsistenzordnung $p = 2$. In jedem Integrationsschritt werden drei Funktionsaufrufe (zwei bez. $f(x, y)$, einer bez. $f_y(x, y)$) benötigt. Wählt man speziell $t_1 = 1$ (vgl. Nickel und Rieder [1]), so stehen mit (2.2) und (2.4) in den Punkten $x_n + h$, $n = 0, 1, \dots, N - 1$, die Funktionswerte $f(x_n + h, \overline{Y}_1)$ und $f(x_n + h, \overline{Y}_2)$ zur Verfügung. Für $\overline{Y}_2 \neq \overline{Y}_1$ kann dann $a$ in den Punkten $x_n + h$ durch

$$a \approx \frac{f(x_n + h, \overline{Y}_2) - f(x_n + h, \overline{Y}_1)}{\overline{Y}_2 - \overline{Y}_1} \tag{2.12}$$

approximiert werden, so daß in $I_n$, $n = 1, 2, \dots, N - 1$, die Funktionsberechnung $f_y(x_n, Y_n)$ entfällt. Die Approximation (2.12) kann jedoch leicht zu numerischen Ungenauigkeiten führen.

*Der lokale Diskretisierungsfehler*

$$\sigma(x; h) = \frac{1}{h} [y(x + h) - \overline{Y}_2]$$

*genügt in* 1. *Näherung der Beziehung*

$$\sigma(x; h) = \left(1 - \frac{3}{2} t_1\right)(D^2 f - aDf) \frac{h^3}{3!}.$$

*Wählt man speziell* $t_1 = 2/3$, *so besitzt* $\overline{Y}_2$ *sogar die Konsistenzordnung* $p = 3$.

3. Ein Vergleich der Koeffizienten der Gleichung (2.10) mit denen gleicher $h$-Potenzen im Ausdruck (2.11) führt mit (2.8) für die Parameter $t_1$, $t_2$ und $t_3$ auf die folgenden Bedingungsgleichungen

$$\text{für } \frac{h^4}{24}: \frac{4}{3}(t_2 + t_3) - 2t_2t_3 = 1,$$

$$\text{für } \frac{h^5}{120}: \frac{5}{3}(t_2^2 + t_2t_3 + t_3^2) - \frac{5}{2}t_2t_3(t_2 + t_3) = 1, \tag{2.13}$$

$$\frac{5}{3}(t_2 + t_3 - 2t_2t_3) = 1,$$

$$\frac{5}{3}(t_2^2 + t_2t_3 + t_3^2)\left(\frac{3}{2}\frac{t_1(t_2 + t_3)}{t_2^2 + t_2t_3 + t_3^2} - 1\right) -$$

$$- \frac{5}{2}t_2t_3(t_2 + t_3)\left(\frac{3}{2}\frac{t_1}{t_2 + t_3} - 1\right) = 0.$$

Mit $t_3 \neq 2/3$ liefert die 1. Gleichung von (2.13)

$$t_2 = \frac{4t_3 - 3}{6t_3 - 4}. \tag{2.14}$$

Für

$$0 < t_3 \leqq \frac{1}{2} \quad \text{bzw.} \quad \frac{3}{4} < t_3 \leqq 1 \tag{2.15}$$

ist dann auch $0 < t_2 \leqq 1$. Folglich ergibt Gleichung (2.9) mit $0 < t_1 \leqq 1$ und (2.14), (2.15) eine Klasse von Einschrittverfahren der Konsistenzordnung $p = 4$. In jedem Diskretisierungsintervall $I_n$ sind dabei fünf Funktionsberechnungen (vier bez. $f(x, y)$, einer bez. $f_y(x, y)$) erforderlich. Im Spezialfall $t_3 = 1$, so daß nach (2.14) $t_2 = 1/2$ gilt (vgl. NICKEL und RIEDER [1]), kann — analog den vorhergehenden Betrachtungen — die Berechnung von $f_y(x_n, Y_n)$ in $I_n$, $n = 1, 2, \ldots, N - 1$, wieder eingespart werden.

*Nach (2.13) gibt es nun genau eine Parameterwahl, für die $\overline{Y}_4$ die Konsistenzordnung $p = 5$ aufweist, nämlich*

$$t_1 = \frac{8}{15}, \; t_2 = \frac{6 - \sqrt{6}}{10} \approx 0.35505102, \; t_3 = \frac{6 + \sqrt{6}}{10} \approx 0.84494898. \tag{2.16}$$

## 3. Zur Stabilität der Einschrittverfahren

Im folgenden soll das Stabilitätsverhalten der 4stufigen Einschrittverfahren (2.9) bez. der von PROTHERO und ROBINSON [3] eingeführten skalaren Anfangswertaufgabe

$$y' = v'(x) + \lambda(y - v(x)), \quad \lambda \in C, \quad \text{Re}(\lambda) < 0,$$

$$y(x_0) = y_0, \tag{3.0}$$

betrachtet werden, wobei $v(x)$ eine für $x \geqq x_0$ gegebene, stetig differenzierbare Funktion mit beschränkter Ableitung sei. Das Testanfangswertproblem (3.0) besitzt die Lösung

$$y(x) = e_0(\lambda(x - x_0))[y_0 - v(x_0)] + v(x),$$

die — unabhängig von der Anfangsbedingung — die Eigenschaft

$$\lim_{x \to \infty} y(x) = v(x)$$

hat. Für die nachstehenden Untersuchungen ist es zweckmäßig, die Verfahren (2.9) in der rekursiven Form

$$Y_{n+1}^{(0)} = Y(x_n),$$

$$Y_{n+1}^{(i)} = Y_{n+1}^{(0)} + h[e_1(a_nt_ih)f_nt_i + e_2(a_nt_ih)\sum_{l=1}^{i-1} a_{il}A_n^{(l)} +$$

$$+ e_3(a_nt_ih)\sum_{l=2}^{i-1} b_{il}A_n^{(l)}], \quad i = 1(1)4, \tag{3.1}$$

$$Y(x_n + h) := Y_{n+1}^{(4)}, \qquad n = 0, 1, \ldots, N - 1,$$

zu schreiben, wobei

$$t_4 = 1; \quad a_n = f_y(x_n, Y(x_n)); \quad f_n = f(x_n, Y(x_n));$$

$$A_n^{(l)} = f(x_n + t_lh, Y_{n+1}^{(l)}) - a_n(Y_{n+1}^{(l)} - Y(x_n)) - f_n,$$

ist, und die Koeffizienten $a_{il}$, $b_{il}$ durch

$$a_{21} = \frac{t_2^2}{t_1}; \quad a_{31} = \frac{t_3^2}{t_1}; \quad a_{32} = a_{41} = 0; \quad a_{42} = \frac{t_3}{t_2(t_3 - t_2)};$$

$$a_{43} = \frac{-t_2}{t_3(t_3 - t_2)}$$

$$b_{32} = 0; \quad b_{42} = -\frac{1}{t_2(t_3 - t_2)}; \quad b_{43} = \frac{1}{t_3(t_3 - t_2)}$$

gegeben sind. Aus (3.1) erhält man bei Anwendung auf (3.0) für die globalen (asymptotischen) Diskretisierungsfehler

$$\varepsilon_{n+1}^{(i)} = Y_{n+1}^{(i)} - v(x_n + t_ih); \quad i = 1(1)4, \quad \text{mit} \quad \varepsilon_{n+1} := \varepsilon_{n+1}^{(4)}$$

die Differenzengleichungen

$$\varepsilon_{n+1}^{(i)} = e_0(t_iz)\,\varepsilon_n + h\xi_n^{(i)}; \quad z = h\lambda, \quad \varepsilon_n := Y_{n+1}^{(0)} - v(x_n), \tag{3.2}$$

wobei die *lokalen* (asymptotischen) *Diskretisierungsfehler* $\xi_n^{(i)}$ durch

$$\xi_n^{(1)} = t_1e_1(t_1z)\,\delta(x_n) + \frac{1}{h}[e_0(t_1z)\,v(x_n) - v(x_n + t_1h)],$$

$$\xi_n^{(i)} = \left[t_ie_1(t_iz) - \frac{t_i^2}{t_1}e_2(t_iz)\right]\delta(x_n) + \frac{t_i^2}{t_1}e_2(t_iz)\,\delta(x_n + t_1h) +$$

$$+ \frac{1}{h}[e_0(t_iz)\,v(x_n) - v(x_n + t_ih)]; \qquad i = 2, 3, \tag{3.3}$$

$$\xi_n := \xi_n^{(4)} = e_1(z)\,\delta(x_n) + \frac{1}{t_3t_2}[e_3(z) - (t_2 + t_3)\,e_2(z)]\,\delta(x_n) +$$

$$+ \frac{1}{t_3 - t_2}\sum_{l=2}^{3} q_l(z)\frac{1}{t_l}\delta(x_n + t_lh) + \frac{1}{h}[e_0(z)\,v(x_n) - v(x_n + h)]$$

mit

$$\delta(x) = v'(x) - \lambda v(x), \; q_2(z) = t_3e_2(z) - e_3(z); \; q_3(z) = e_3(z) - t_2e_2(z)$$

gegeben sind. Aus (3.3) ergibt sich

$$\xi_n^{(1)} = [e_1(t_1z)\,v'(x_n) - v'(x_n + \theta_1t_1h)]\,t_1; \quad 0 < \theta_1 < 1,$$

$$\xi_n^{(i)} = e_2(t_iz)\,[v'(x_n + t_1h) - v'(x_n)]\frac{t_i^2}{t_1} + e_1(t_iz)\,v'(x_n)\,t_i -$$

$$- [(e_1(t_iz) - 1)\,v'(x_n + \theta_1t_1h) + v'(x_n + \theta_it_ih)]\,t_i; \tag{3.4}$$

$$0 < \theta_i < 1; \quad i = 2, 3,$$

$$\xi_n^{(4)} = \left[e_1(z) + \frac{1}{t_2t_3}(e_3(z) - (t_2 + t_3)\,e_2(z))\right]v'(x_n) +$$

$$+ \frac{1}{t_3 - t_2}\sum_{l=2}^{3} q_l(z)\left[\frac{1}{t_2}v'(x_n + t_lh) - zv'(x_n + \theta_lt_lh)\right] -$$

$$- v'(x_n + \theta_4h); \qquad 0 < \theta_4 < 1.$$

Die lokalen Diskretisierungsfehler $\xi_n^{(i)}$, $i = 1(1)4$, sind folglich für jedes feste $\overline{h} > 0$ auf der Menge $R := \{(h, z) \mid \text{Re}(z) < 0, h \in (0, \overline{h}]\}$ gleichmäßig beschränkt. Für alle $(h, z)$ mit $h \in (0, \overline{h}]$ und $\text{Re}(z) \leqq \lambda_0$, $\lambda_0 < 0$, gilt dann

$$\frac{|\xi_n^{(i)}|}{1 - |e_0(t_iz)|} < K_i, \quad i = 1(1)4.$$

Weiterhin folgt aus (3.2)

$$|\varepsilon_{n+1}^{(i)}| \leqq |e_0(t_iz)|\,|\varepsilon| + h|\xi_n^{(i)}|, \quad i = 1(1)4,$$

$$= (1 - |e_0(t_iz)|)\frac{h\,|\xi_n^{(i)}|}{1 - |e_0(t_iz)|} + |e_0(t_iz)|\,|\varepsilon_n|.$$

Daraus folgt wegen $1 - |e_0(t_iz)| > 0$

$$|\varepsilon_{n+1}^{(i)}| \leqq \max\left(|\varepsilon_n|, \frac{h|\xi_n^{(i)}|}{1 - |e_0(t_iz)|}\right),$$

so daß die *globalen Diskretisierungsfehler* $\varepsilon_n^{(i)}$, $i = 1(1)4$, mit $n$ für alle $h \in (0, \overline{h}]$ und alle $\lambda$ mit $\overline{h} \cdot \text{Re}(\lambda) \leqq \lambda_0$ gleichmäßig beschränkt sind. Die Einschrittverfahren (2.9) sind somit gemäß VERWER [5] intern $S$-stabil. *Für die speziellen Parameter*

$$0 < t_1 \leqq 1, \quad t_2 = 1/2, \quad t_3 = 1 \tag{3.5}$$

*folgt aus* (3.4) *unter Benutzung von* (2.0) *für jede Schrittweite* $h > 0$

$$\lim_{\text{Re}(z) \to -\infty} \xi_n = 0,$$

*so daß für* (3.5) *die Verfahren* (2.9) *gemäß* PROTHERO *und* ROBINSON [3] *„stiff" genau sind.*

In analoger Weise zeigt man, daß das einstufige Verfahren (2.2) ($t_1 = 1$) $S$-stabil (vgl. VERWER [4]) und das zweistufige Verfahren (2.4) für $0 < t_1 \leqq 1$ und $t_2 = 1$ intern $S$-stabil ist. Wählt man $t_1 = 1$, so ist das Verfahren (2.4) „stiff" genau.

Literatur


1 NICKEL, K.; RIEDER, P., Ein neues RUNGE-KUTTA-ähnliches Verfahren, ISNM, Vol. 9, Numerische Mathematik, Differentialgleichungen, Approximationstheorie, Basel: Birkhäuser 1968, S. 83–96.
2 GRIGORIEFF, R. D., Numerik gewöhnlicher Differentialgleichungen, Band 1, B. G. Teubner, Stuttgart 1972.
3 PROTHERO, A.; ROBINSON, A., On the stability and accuracy of one-step methods for solving stiff systems of ordinary differential equations, Math. Comput. 28, No. 125, 145–162 (1974).
4 VERWER, J. G., S-stability properties for generalized RUNGE-KUTTA methods, Numer. Math. 27, 359–370 (1977).
5 VERWER, J. G., Internal S-stability for generalized RUNGE-KUTTA methods, Report NW 21/75, Mathematisch Centrum, Amsterdam 1975.


Eingereicht: 2. 10. 1978

*Anschrift*: Doz. Dr. sc. KARL STREHMEL, Martin-Luther-Universität, Sektion Mathematik, DDR 402 Halle/S., Weinbergweg 17

ZAMM 60, 53–54 (1980)

K. C. GARG/K. SWARUP

## The Use of Cuts in Linear Fractional Functional Complementary Programming

*A Complementary Programming Problem with Linear Fractional Objective Function* (CPPLFOF) is a Linear Fractional Functional Programming Problem (LFFPP) namely;

Maximize:

$$z = \frac{dx + eu + fv + g}{\alpha x + \beta u + \gamma v + \delta} \tag{P}$$

subject to

$$Ax + Bu + Cv \leqq h, \; x, u, v \geqq 0$$

with the additional restriction that

$$uv = 0 ,$$

known as *complementarity condition*. To solve a CPPLFOF, a branch and bound algorithm was proposed in [1]. In this paper, a new type of constraints derived from the complementarity condition, $uv = 0$, are discussed. These constraints are called *C-cuts* (complementarity cuts) and possess similar properties as proposed by GOMORY [2, 3] and KELLEY [7]. One of the applications of C-cuts is that they can be used in the branch and bound algorithm proposed by the authors in [1] to make the computational work easier.

Consider a linear fractional functional programming problem ($\bar{\text{P}}$) i.e. problem (P) without the complementarity condition, $uv = 0$. Let $C^*$ be a given set of pairs of variables of the type $(u_j, v_j)$, $1 \leqq j \leqq m$. A CPPLFOF is a LFFPP($\bar{\text{P}}$) with the additional restriction $uv = 0$ for all $(u_j, v_j) \in C^*$, $1 \leqq j \leqq m$, at least one of $u_j$ and $v_j$ is zero for all $(u_j, v_j) \in C^*$, $1 \leqq j \leqq m$. Let $t = (x, u, v)$ so that $t$ is a vector of variables of problem (P) or ($\bar{\text{P}}$) having $n + 2m$ components.

For each maximization problem Q ($= \bar{\text{P}}$ or P or anything else), a solution $t$ is *feasible* if it satisfies Q's constraints. Here $t = (x_1, \dots, x_n; u_1, \dots, u_m; v_1, \dots, v_m)$. The value of the objective function at $t$ is denoted by $z(t)$. $F(\text{Q})$ denotes the *set of all feasible solutions* of Q. $\bar{t} \in F(\text{Q})$ is an *optimal solution* of Q satisfying $z(\bar{t}) \geqq z(t)$ for all $t \in F(\text{Q})$. The *value of the optimal solution* of Q is denoted by $z(\text{Q})$.

Let Q be a LFFPP having the same objective function as (P) and satisfying $F(\text{P}) \subset F(\text{Q})$. Let $\bar{t}$ be an optimal solution of Q. Let $w_1, w_2, \dots, w_{n+2m}$ be the nonbasic variables of problem Q. Then each variable $t_k$ can be written in terms of these, namely

$$t_k = \alpha_{k0} + \sum_{j=1}^{n+2m} \alpha_{kj}(-w_j) , \; k = 1, \dots, n + 2m + p ,$$

$$w_j \geqq 0 , \qquad j = 1, \dots, n + 2m .$$

Similarly the objective function can also be written in terms of the nonbasic variables $w_1, \dots, w_{n+2m}$. $\bar{t}$ is obtained by setting each $w_j$ to zero. In particular, $\alpha_{k0} \geqq 0$, $k = 1, \dots, n + 2m + p$.

Let us consider a constraint given by

$$s = \gamma_0 + \sum_{j=1}^{n+2m} \gamma_j(-w_j), \; s \geqq 0 . \tag{1}$$

This constraint is known as a C-cut with respect to $\bar{t}$ if it does not exclude any point in $F(\text{P})$ but does exclude $\bar{t}$ (i. e. $\gamma_0 < 0$). Now consider problem ($\bar{\text{P}}$) and $F(\text{P}) \subset F(\bar{\text{P}})$. Let $w_j$ ($j = 1, \dots, n + 2m$) be the nonbasic variables yielding an optimal solution $\bar{t}$ of $\bar{\text{P}}$. Let there exist a pair of variables $(u_i, v_i) \in C^*$ such that

$$u_i = \overset{u}{\alpha}_{i0} + \sum_{j=1}^{n+2m} \overset{u}{\alpha}_{ij} (- w_j), \quad v_i = \overset{v}{\alpha}_{i0} + \sum_{j=1}^{n+2m} \overset{v}{\alpha}_{ij}(- w_j) \tag{2}$$

and

$$\overset{u}{\alpha}_{i0} > 0 , \quad \overset{v}{\alpha}_{i0} > 0 , \tag{3}$$

i.e. $u_i v_i = 0$ is not satisfied by the optimal solution $\bar{t}$ obtained by setting each $w_j$ to zero. Thus the constraint (1), whose coefficients are

$$\gamma_0 = -1 ,$$

$$\gamma_j = 0 \text{ if } \alpha_{lj} \leqq 0, \, l \text{ corresponding to } u_i, v_i$$

$$= \text{Min} \left\{ \begin{array}{l} -(\alpha_{lj})/\alpha_{l0} \colon \alpha_{lj}/\alpha_{l0} > 0, l \\ \text{corresponding to } u_i, v_i; \, j = 1, \dots, n + 2m \end{array} \right\} \tag{4}$$

is the required C-cut as will be proved in the following theorem based on similar lines as in [5, 6].

Theorem: *Let* $\bar{\text{P}}$, $\bar{t}$, $w_j$ ($j = 1, \dots, n + 2m$) *be defined as usual. If there exists a pair of variables* $(u_i, v_i) \in C^*$ *satisfying* (2) *and* (3), *the constraints defined by* (1) *and* (4) *exclude* $\bar{t}$ *from* $F(\bar{\text{P}})$ *but don't exclude any point in* $F(\text{P})$.

Proof: To prove the first part of the theorem, we shall show that the constraints defined by (1) and (4) exclude $\bar{t}$ from the set of all feasible solutions of problem $\bar{\text{P}}$. Since $\bar{t}$ is obtained by setting each $w_j$ to zero, therefore from (1) follows $s = -1 \ngeqq 0$ which means that the C-cut excludes $\bar{t}$ from $F(\bar{\text{P}})$.

To prove the second part of the theorem, we shall show that no point $(w_1, \dots, w_{n+2m})$ excluded by the C-cut satisfies $u_i v_i = 0$ where $u_i$ and $v_i$ can be calculated from $(w_1, \dots, w_{n+2m})$ using (2). Obviously (1) and (4) give

$$\sum_{j=1}^{n+2m} \gamma_j(- w_j) < 1 , \quad w_j \geqq 0 , \tag{5}$$

defining the region of $w$'s that is cut off by the C-cut (1) and (4). Since $\overset{u}{\alpha}_{i0} > 0$, therefore from (5) follows

$$\overset{u}{\alpha}_{i0} \sum_{j=1}^{n+2m} \gamma_j(-w_j) < \overset{u}{\alpha}_{i0} ,$$

i.e.

$$\overset{u}{\alpha}_{i0} - \overset{u}{\alpha}_{i0} \sum_{j=1}^{n+2m} \gamma_j(-w_j) > 0 . \tag{6}$$

By definition of $\gamma_j$, $\overset{u}{\alpha}_{ij} > 0$, whence follows

$$\gamma_j \leqq -\frac{\overset{u}{\alpha}_{ij}}{\overset{u}{\alpha}_{i0}} \text{ and } \overset{u}{\alpha}_{ij} \leqq -\overset{u}{\alpha}_{i0}\gamma_j .$$

Consequently from $\overset{u}{\alpha}_{ij} \leqq 0$ follows $-\overset{u}{\alpha}_{i0} \gamma_j \geqq 0$. Therefore

$$u_i = \overset{u}{\alpha}_{i0} + \sum_{j=1}^{n+2m} \overset{u}{\alpha}_{ij}(-w_j) \geqq \overset{u}{\alpha}_{i0} + \sum_{j=1}^{n+2m} \overset{u}{\alpha}_{i0}\gamma_j w_j$$

$$= \overset{u}{\alpha}_{i0} - \sum_{j=1}^{n+2m} \overset{u}{\alpha}_{i0} \gamma_j(-w_j) > 0 , \quad \text{from (6).}$$

In a similar fashion, it can be shown that $v_i > 0$ for any $w$ in the feasible region of (5). Thus we conclude that any point excluded from $F(\overline{\mathrm{P}})$ by (1) satisfies $u_i v_i \neq 0$. Q.E.D.

Now after adding the C-cut (1) to problem $(\overline{\mathrm{P}})$, suppose we get a LFFPP $\overline{\mathrm{P}}_1$. Then problem $\overline{\mathrm{P}}_1$ is solved with the help of the technique developed by SWARUP [9]. If again there exists a pair of variables $(u_i, v_i) \in C^*$ satisfying (2) and (3), we obtain a new C-cut. Adding this C-cut to problem $\overline{\mathrm{P}}_1$, we get the LFFPP $\overline{\mathrm{P}}_2$ and solve it again like $\overline{\mathrm{P}}_1$. Repeating this process, we get problem $\overline{\mathrm{P}}_k$ which is problem $\overline{\mathrm{P}}$ with $k$ C-cuts added. It follows that (from the theorem proved above) if $k_1 \leqq k_2$, then

$$F(\overline{\mathrm{P}}_{k_1}) \supset F(\overline{\mathrm{P}}_{k_2}) \supset \ldots \supset F(\overline{\mathrm{P}}) \supset F(\mathrm{P})$$

and this means that if $k_1 \leqq k_2$, then

$$z(\overline{\mathrm{P}}_{k_1}) \geqq z(\overline{\mathrm{P}}_{k_2}) \geqq \ldots \geqq z(\overline{\mathrm{P}}) \geqq z(\mathrm{P}).$$

If we could specify the generation method of C-cuts such that $z(\overline{\mathrm{P}}_k) = z(\mathrm{P})$ and $\bar{t}_k \in F(\mathrm{P})$ where $\bar{t}_k$ is an optimal solution of $\overline{\mathrm{P}}_k$, $k$ being finite, then the resulting algorithm is a cutting plane method. Finite convergence of such an algorithm is guaranteed if we introduce another type of cuts known as $\Delta$-cuts [5]. Thus this paper is an attempt to make use of C-cuts in linear fractional functional complementary programming.

## Numerical example

Consider the following complementary programming problem with linear fractional objective function:

Maximize:

$$z = \frac{2x_1 + x_2}{4x_1 + x_2 + 1} \qquad \text{(P)}$$

subject to

$$-2x_1 + x_2 \leqq 1,$$
$$2x_1 + 5x_2 \leqq 23,$$
$$2x_1 + x_2 \leqq 15,$$
$$x_1 - 2x_2 \leqq 5,$$
$$x_1 \cdot x_2 = 0, \quad x_1, x_2 \geqq 0\,.$$

Problem $(\overline{\mathrm{P}})$ is problem (P) without the complementarity condition and thus is an ordinary LFFPP. Solving $(\overline{\mathrm{P}})$ using the simplex technique of linear fractional functional programming [8], we get an optimal solution given by

$x_1 = 3/2$, $x_2 = 4$; $z = 7/11$ (Tableau 1).

Obviously, the optimal solution of problem $(\overline{\mathrm{P}})$ is not an optima solution of problem (P) since the complementarity condition $x_1 x_2 = 0$, is not satisfied. Therefore introducing a C-cut (1) represented by

$$s_1 = -1 + 1/24\, x_3 + 1/18\, x_4$$

with the help of (4) and tableau 1, we get tableau 2.

Tableau 1

| | | | $c_j$ | 2 | 1 | 0 | 0 | 0 | 0 |
|---|---|---|---|---|---|---|---|---|---|
| | | | $d_j$ | 4 | 1 | 0 | 0 | 0 | 0 |
| $D_H$ | $C_H$ | Variables in basis | $x_H$ | $x_1$ | $x_2$ | $x_3$ | $x_4$ | $x_5$ | $x_6$ |
| 4 | 2 | $x_1$ | 3/2 | 1 | 0 | −5/12 | 1/12 | 0 | 0 |
| 1 | 1 | $x_2$ | 4 | 0 | 1 | 1/6 | 1/6 | 0 | 0 |
| 0 | 0 | $x_5$ | 8 | 0 | 0 | 2/3 | −1/3 | 1 | 0 |
| 0 | 0 | $x_6$ | 23/2 | 0 | 0 | 3/4 | 1/4 | 0 | 1 |
| $z^{(2)} = 11$ | | | $c_j - z_j^{(1)}$ | 0 | 0 | 2/3 | −1/3 | 0 | 0 |
| $z^{(1)} = 7$ | | | $d_j - z_j^{(2)}$ | 0 | 0 | 3/2 | −1/2 | 0 | 0 |
| $z = 7/11$ | | | $\Delta_j$ | 0 | 0 | −19/6 | −1/6 | 0 | 0 |

Since $s_1 < 0$ and all $\Delta_j \leqq 0$, apply the technique of linear fractional functional programming developed by SWARUP [9]. Thus to derive the variable $s_1$ out of basis, a pivot operation is performed on the starred element of tableau 2, we get tableau 3.

Tableau 2

| | | | $c_j$ | 2 | 1 | 0 | 0 | 0 | 0 | 0 |
|---|---|---|---|---|---|---|---|---|---|---|
| | | | $d_j$ | 4 | 1 | 0 | 0 | 0 | 0 | 0 |
| $D_H$ | $C_H$ | Variables in basis | $x_H$ | $x_1$ | $x_2$ | $x_3$ | $x_0$ | $x_5$ | $x_6$ | $s_1$ |
| 4 | 2 | $x_1$ | 3/2 | 1 | 0 | −5/12 | 1/12 | 0 | 0 | 0 |
| 1 | 1 | $x_2$ | 4 | 0 | 1 | 1/6 | 1/6 | 0 | 0 | 0 |
| 0 | 0 | $x_5$ | 8 | 0 | 0 | 2/3 | −1/3 | 1 | 0 | 0 |
| 0 | 0 | $x_6$ | 23/2 | 0 | 0 | 3/4 | 1/4 | 0 | 1 | 0 |
| 0 | 0 | $s_1$ | −1 | 0 | 0 | −1/24 | −1/18* | 0 | 0 | 1 |
| $z^{(2)} = 11$ | | | $c_j - z_j^{(1)}$ | 0 | 0 | 2/3 | −1/3 | 0 | 0 | 0 |
| $z^{(1)} = 7$ | | | $d_j - z_j^{(2)}$ | 0 | 0 | 3/2 | −1/2 | 0 | 0 | 0 |
| $z = 7/11$ | | | $\Delta_j$ | 0 | 0 | −19/6 | −1/6 | 0 | 0 | 0 |
| | | | $\delta_j$ | 0 | 0 | −47/264 | −1/99 | 0 | 0 | 0 |
| | | | $\Delta_j/\delta_j$ | 0 | 0 | 836/47 | 99/6 | 0 | 0 | 0 |

From tableau 3, we note that the complementarity condition, $x_1 x_2 = 0$, is satisfied. Hence the optimal solution to problem (P) is given by

$x_1 = 0$, $x_2 = 1$; $z = 1/2$.

Therefore the procedure terminates here.

Tableau 3

| | | | $c_j$ | 2 | 1 | 0 | 0 | 0 | 0 | 0 |
|---|---|---|---|---|---|---|---|---|---|---|
| | | | $d_j$ | 4 | 1 | 0 | 0 | 0 | 0 | 0 |
| $D_H$ | $C_H$ | Variables in basis | $x_H$ | $x_1$ | $x_2$ | $x_3$ | $x_4$ | $x_5$ | $x_6$ | $s_1$ |
| 4 | 2 | $x_1$ | 0 | 1 | 0 | −23/48 | 0 | 0 | 0 | 3/2 |
| 1 | 1 | $x_2$ | 1 | 0 | 1 | 1/24 | 0 | 0 | 0 | 3 |
| 0 | 0 | $x_5$ | 14 | 0 | 0 | 11/12 | 0 | 1 | 0 | −6 |
| 0 | 0 | $x_6$ | 7 | 0 | 0 | 9/16 | 0 | 0 | 1 | 9/2 |
| 0 | 0 | $x_4$ | 18 | 0 | 0 | 3/4 | 1 | 0 | 0 | −18 |
| $z^{(2)} = 2$ | | | $c_j - z_j^{(1)}$ | 0 | 0 | 11/12 | 0 | 0 | 0 | −6 |
| $z^{(1)} = 1$ | | | $d_j - z_j^{(2)}$ | 0 | 0 | 45/24 | 0 | 0 | 0 | −9 |
| $z = 1/2$ | | | $\Delta_j$ | 0 | 0 | −1/24 | 0 | 0 | 0 | −3 |

## Acknowledgement

I am highly grateful to the expert, who refereed this paper, for his opinion regarding this paper and valuable suggestions which encouraged and helped us in revising this manuscript.

## References

1 GARG, K. C.; SWARUP, KANTI, Complementary programming with linear fractional objective function, CCERO (Belgium) **20**, 83–94 (1978).
2 GOMORY, R. E., An algorithm for the mixed-integer problem, P-1885, The Rand Corporation, Inc. 1960.
3 GOMORY, R. E., An algorithm for the integer solutions to linear programs, in Recent Advances in Mathematical Programming, R. L. GRAVES, P. WOLFE (eds.), Mc-Graw Hill, New York 1963.
4 IBARAKI, T., Complementary programming, Operations Res. **19**, 1523–1528 (1971).
5 IBARAKI, T., Complementary convex programming, J. Operational Res. Soc. Japan **15**, 138–160 (1972).
6 IBARAKI, T., The use of cuts in complementary programming, Operations Res. **21**, 353–359 (1973).
7 KELLEY, J. E. Jr., The cutting plane method for solving convex programs, J. Soc. Industr. Appl. Maths. 8, 703–712 (1960).
8 SWARUP, K., Linear fractional functional programming, Operations Res. **13**, 1029–1036 (1965).
9 SWARUP, K., Some aspects of linear fractional functional programming Australian J. Statistics **7**, 90–104 (1965).

Eingereicht: 20. 5. 78, revidierte Fassung 26. 9. 1978

*Anschriften*: Dr. K. C. GARG, Department of Mathematics, NREC College, Khurja-203131; Professor Dr. KANTI SWARUP, Department of Operations Research, Indian Institute of Public Administration, Indaprastha Estate, Ring Road, New Delhi-110002, India

ZAMM 60, 55 (1980)

E. Kreyszig

## Newton-Cotes-Koeffizienten und Binomialverteilung

Zwischen den Newton-Cotes-Integrationsformeln [2, S. 482] und der Binomialverteilung [1, S. 107] besteht ein interessanter Zusammenhang: Wir zeigen, daß sich die Koeffizienten dieser Formeln in einfacher Weise aus einer Matrix gewinnen lassen, die man der genannten Verteilung zuordnen kann. In der Tat gilt der folgende

Satz: *Der Koeffizient* $a_k^{(n)}$, $k = 0, \dots, n$, *in der Newton-Cotes-Formel*

$$\int_0^1 f(x)\,dx \approx \frac{1}{n+1} \sum_{k=0}^{n} a_k^{(n)} f\left(\frac{k}{n}\right)$$

*ist die Summe der* $(k+1)$*-ten Spalte der Matrix* $B_{(n)}^{-1}$. *Hierbei ist* $B_{(n)} = (p_{jk}^{(n)})$ *die* $(n+1)$*-reihige quadratische Matrix mit den Elementen*

$$p_{jk}^{(n)} = n^{-n}\binom{n}{k} j^k (n-j)^{n-k}, \qquad j, k = 0, \dots, n. \tag{1}$$

Beweis: Es sei $n \in \mathbb{N}$ fest. $g(x)$ sei eine auf $[0, 1]$ definierte Funktion. Der Spaltenvektor $g = (g_k)$, $g_k = g(k/n)$, $k = 0, \dots, n$, heiße der *zugeordnete Vektor* von $g(x)$, und

$$B_n(g; x) = \sum_{k=0}^{n} g\left(\frac{k}{n}\right)\binom{n}{k} x^k (1-x)^{n-k} \tag{2}$$

heiße das *zugeordnete Bernstein-Polynom* von $g(x)$. Der zugeordnete Vektor von $B_n(g; x)$ ist dann $B_{(n)}g$ mit $B_{(n)} = (p_{jk}^{(n)})$ gemäß (1). Die Matrix $B_{(n)}$ ist nichtsingulär. Aus (2) folgt durch Integration und Vereinfachung

$$\int_0^1 B_n(g; x)\,dx = \frac{1}{n+1} \sum_{k=0}^{n} g\left(\frac{k}{n}\right). \tag{3}$$

Es sei $f(x)$ das Polynom vom Grade $\leqq n$, das durch $n+1$ gegebene Werte $f(k/n)$, $k = 0, \dots, n$, eindeutig bestimmt ist. Dann gilt $f = B_{(n)}g$ mit

$$g = B_{(n)}^{-1} f. \tag{4}$$

Es sei $B_{(n)}^{-1} = (b_{jk})$. Aus (3) und (4) folgt

$$\int_0^1 f(x)\,dx = \frac{1}{n+1} \sum_{j=0}^{n} \sum_{k=0}^{n} b_{jk} f\left(\frac{k}{n}\right).$$

Damit ist der Satz bewiesen.

Beispielsweise ist für $n = 2$ (Simpson-Regel)

$$B_{(2)} = \frac{1}{4}\begin{bmatrix} 4 & 0 & 0 \\ 1 & 2 & 1 \\ 0 & 0 & 4 \end{bmatrix}, \qquad B_{(2)}^{-1} = \frac{1}{2}\begin{bmatrix} 2 & 0 & 0 \\ -1 & 4 & -1 \\ 0 & 0 & 2 \end{bmatrix}$$

und für $n = 3$ (Drei-Achtel-Regel)

$$B_{(3)} = \frac{1}{27}\begin{bmatrix} 27 & 0 & 0 & 0 \\ 8 & 12 & 6 & 1 \\ 1 & 6 & 12 & 8 \\ 0 & 0 & 0 & 27 \end{bmatrix},$$

$$B_{(3)}^{-1} = \frac{1}{6}\begin{bmatrix} 6 & 0 & 0 & 0 \\ -5 & 18 & -9 & 2 \\ 2 & -9 & 18 & -5 \\ 0 & 0 & 0 & 6 \end{bmatrix}$$

usw.

Literatur


1 Kreyszig, E., Statistische Methoden und ihre Anwendungen, 7. Aufl., Vandenhoeck und Ruprecht, Göttingen 1979.
2 Todd, J. (Ed.), Survey of Numerical Analysis, McGraw-Hill, New York 1962.


Eingereicht: 31. 5. 1979

*Anschrift*: Prof. Dr. E. Kreyszig, Department of Mathematics, University of Windsor, Windsor, Ont. N9B 3P4 Canada

ZAMM 60, 55–56 (1980)

L. K. Roy

## More about Entropies of Power Distributions

Introduction

Following the notation of Mitter and Mathur [2] and Roy [3] we may define the Shannon *entropy* of the power distribution $P^\beta = (p_1^\beta, p_2^\beta, \dots, p_n^\beta)$ as

$$H(P^\beta) = -\frac{\sum_{i=1}^{n} p_i^\beta \log p_i^\beta}{\sum_{i=1}^{n} p_i^\beta} \tag{1}$$

for $\sum_{i=1}^{n} p_i^\beta \leqq 1$, and the Rényi *entropy of order* $\alpha$ of the power distribution $P^\beta$ is defined as

$$H_\alpha(P^\beta) = \frac{1}{1-\alpha} \log \left[\frac{\sum_{i=1}^{n} p_i^{\alpha\beta}}{\sum_{i=1}^{n} p_i^\beta}\right] \tag{2}$$

for $\sum_{i=1}^{n} p_i^\beta \leqq 1$.

The main result of Mitter and Mathur [2] was

$$\frac{1}{\gamma} H(P^\gamma) \geqq \frac{1}{\beta} H(P^\beta) \quad \text{for } \beta \geq \gamma \tag{3}$$

while Roy [3] proved that if $\sum_{i=1}^{n} p_i^\beta = 1$, then

$$\frac{1}{\alpha} H(P^{\alpha\beta}) \leqq H_\alpha(P^\beta) \quad \text{for } \alpha \geqq 1. \tag{4}$$

The following results indicate the relationship between $H$ and $H_\alpha$.

Theorems:

Theorem 1: *If* $\sum_{i=1}^{n} p_i^{\gamma\beta} = 1$, $0 < \gamma \leqq 2$, $0 < \beta \leqq 2$ *then*

$$\frac{4}{\gamma\beta} H^2(P^{\gamma\beta}) \leqq H(P^{2\gamma}) H(P^{2\beta}). \tag{5}$$

Proof: From Beckenbach and Bellman ([1] page 45) we have the inequality

$$\left(\sum_{i=1}^{n} x_i y_i w_i\right)^2 \leqq \left(\sum_{i=1}^{n} x_i^2 w_i\right)\left(\sum_{i=1}^{n} y_i^2 w_i\right) \tag{6}$$

as long as $x_i > 0$, $y_i > 0$, $w_i \geqq 0$ for $i = 1, 2, \dots, n$ and not all $w_i$ are zero. In (6) let $x_i = p_i^\gamma$, $y_i = p_i^\beta$, $w_i = -\log p_i$. Then

$$\left(-\sum_{i=1}^{n} p_i^\gamma p_i^\beta \log p_i\right)^2 \leqq \left(-\sum_{i=1}^{n} p_i^{2\gamma} \log p_i\right)\left(-\sum_{i=1}^{n} p_i^{2\beta} \log p_i\right),$$

$$\left(-\frac{1}{\gamma\beta} \sum_{i=1}^{n} p_i^{\gamma\beta} \log p_i^{\gamma\beta}\right)^2 \leqq \left(-\frac{1}{2\gamma} \sum_{i=1}^{n} p_i^{2\gamma} \log p_i^{2\gamma}\right) \times$$

$$\times \left(-\frac{1}{2\beta} \sum_{i=1}^{n} p_i^{2\beta} \log p_i^{2\beta}\right),$$

$$\frac{1}{(\gamma\beta)^2}\left[\frac{\sum_{i=1}^{n} p_i^{\gamma\beta}\log p_i^{\gamma\beta}}{\sum_{i=1}^{n} p_i^{\gamma\beta}}\right]^2 \leqq \frac{1}{2\gamma}\left[\frac{\sum_{i=1}^{n} p_i^{2\gamma}\log p_i^{2\gamma}}{\sum_{i=1}^{n} p_i^{2\gamma}}\right]\times$$

$$\times\frac{1}{2\beta}\left[\frac{\sum_{i=1}^{n} p_i^{2\beta}\log p_i^{2\beta}}{\sum_{i=1}^{n} p_i^{2\beta}}\right].$$

Note that in the above expression, since $\sum_{i=1}^{n} p_i^{\gamma\beta} = 1$, and since $0 < \gamma \leqq 2$, $0 < \beta \leqq 2$, it follows that $0 < \gamma\beta \leq 2\beta$ and $0 < \gamma\beta \leq 2\gamma$, so that $\sum_{i=1}^{n} p_i^{2\beta} \leqq 1$ and $\sum_{i=1}^{n} p_i^{2\gamma} \leqq 1$. Hence

$$\frac{1}{(\gamma\beta)^2} H^2(P^{\gamma\beta}) \leqq \frac{1}{4\gamma\beta} H(P^{2\gamma})\, H(P^{2\beta})$$

and (5) is proved.

Corollary: *If* $\sum_{i=1}^{n} p_i = 1$, *then*

$$2H(P) \leq H(P^2) \leqq 2H_2(P)\,. \tag{7}$$

Proof: In (5) let $\gamma = \beta = 1$, and in (4) let $\alpha = 2$ and $\beta = 1$.

Theorem 2: *For* $\alpha > 1$ *and* $\sum_{i=1}^{n} p_i^{\beta} = 1$, $H_\alpha(P^\beta)$ *is a non-increasing function of* $\alpha$.

Proof: In (2) assume that the log has base $D$. Then changing to natural logarithms we get

$$H_\alpha(P^\beta) = \frac{1}{1-\alpha}\frac{1}{\ln D}\ln\left[\frac{\sum_{i=1}^{n} p_i^{\alpha\beta}}{\sum_{i=1}^{n} p_i^{\beta}}\right].$$

So

$$\frac{\partial}{\partial\alpha} H_\alpha(P^\beta) =$$

$$-\frac{1}{\ln D}\frac{1}{(1-\alpha)^2}\ln\left[\frac{\sum_{i=1}^{n} p_i^{\alpha\beta}}{\sum_{i=1}^{n} p_i^{\beta}}\right] + \frac{1}{\ln D}\frac{1}{1-\alpha}\frac{\sum_{i=1}^{n} p_i^{\alpha\beta}\ln p_i^{\beta}}{\sum_{i=1}^{n} p_i^{\alpha\beta}}$$

That is

$$\frac{\partial}{\partial\alpha} H_\alpha(P^\beta) = \frac{1}{(1-\alpha)^2}\log\left[\frac{\sum_{i=1}^{n} p_i^{\alpha\beta}}{\sum_{i=1}^{n} p_i^{\beta}}\right] + \frac{1}{1-\alpha}\frac{\sum_{i=1}^{n} p_i^{\alpha\beta}\log p_i^{\beta}}{\sum_{i=1}^{n} p_i^{\alpha\beta}}$$

after changing back to logarithms with base $D$. Consequently

$$\frac{\partial}{\partial\alpha} H_\alpha(P^\beta) = \frac{1}{1-\alpha} H_\alpha(P^\beta) + \frac{1}{(1-\alpha)\alpha}\frac{\sum_{i=1}^{n} p_i^{\alpha\beta}\log p_i^{\alpha\beta}}{\sum_{i=1}^{n} p_i^{\alpha\beta}}$$

$$= -\frac{1}{\alpha-1} H_\alpha(P^\beta) + \frac{1}{\alpha(\alpha-1)} H(P^{\alpha\beta})\,.$$

Using (4) we get

$$\frac{\partial}{\partial\alpha} H_\alpha(P^\beta) \leq -\frac{1}{\alpha-1} H_\alpha(P^\beta) + \frac{1}{\alpha-1} H_\alpha(P^\beta) = 0$$

which proves the theorem.

Theorem 3: *If* $\alpha > 1$ *and* $\sum_{i=1}^{n} p_i^{\beta} = 1$, *then*

$$H(P^\beta) \geqq H_\alpha(P^\beta)\,. \tag{8}$$

Proof: From Beckenbach and Bellman ([1] page 13) we have the inequality

$$\prod_{i=1}^{n} x_i^{\alpha_i} \leq \sum_{i=1}^{n} \alpha_i x_i \tag{9}$$

for $x_i > 0$, $\alpha_i > 0$, $\sum_{i=1}^{n} \alpha_i = 1$. Taking the logarithm of both sides of (9), substituting $\alpha_i = p_i^{\beta}$ and $x_i = p_i^{(\alpha-1)\beta}$, and multiplying through by $\frac{1}{1-\alpha}$ gives

$$\frac{1}{1-\alpha}\sum_{i=1}^{n} p_i^{\beta}\log p_i^{(\alpha-1)\beta} \geqq \frac{1}{1-\alpha}\log\sum_{i=1}^{n} p_i^{\alpha\beta}\,,$$

$$-\sum_{i=1}^{n} p_i^{\beta}\log p_i^{\beta} \geq \frac{1}{1-\alpha}\log\sum_{i=1}^{n} p_i^{\alpha\beta}$$

which proves the theorem.

Note that combining (8) with (4) produces

$$\frac{1}{\alpha} H(P^{\alpha\beta}) \leqq H(P^\beta) \tag{10}$$

which is another form of (3).

### References


1 Beckenbach, E. F.; Bellman, R., Inequalities, Springer-Verlag, Berlin-Heidelberg-New York 1965.
2 Mitter, J.; Mathur, Y. D., Comparison of entropies of power distributions, ZAMM 52 (1972), 239–240.
3 Roy, L. K., Comparison of Rényi entropies of power distributions, ZAMM 54 (1976), 217–218


Eingereicht: 15. 5. 1979

*Anschrift*: Prof. L. K. Roy, Department of Mathematical Sciences, Lakehead University, Thunder Bay, Ontario, Canada

ZAMM 60, 56–57 (1980)

K. K. Tam

## A Remark on the Brunt-Väisälä Frequency

In studying internal waves in an incompressible, stably stratified fluid, the two-dimensional theory leads to the eigenvalue problem (see Yih 1965)

$$(\bar{\varrho}w')' - \left(k^2\bar{\varrho} + \frac{g}{c^2}\bar{\varrho}'\right) w = 0\,, \qquad ' = \frac{\mathrm{d}}{\mathrm{d}y} \tag{1}$$

with either

$$w(0) = w(d) = 0 \tag{2}$$

for a fixed boundary at $y = d$, or

$$w(0) = 0\,, \quad w'(d) - \frac{g}{c^2} w(d) = 0 \tag{3}$$

for a free boundary at $y = d$. Here $\bar{\varrho}(y)$ is the fluid density, $g$ the gravitational constant, $w(y)$, $k$, $c$ come from the assumed form of the stream function for the internal wave:

$$\psi(x, y, t) = w(y)\, e^{ik(x-ct)}\,.$$

Thus, $w(y)$ is the amplitude of the travelling wave, $k$ the wave number and $c$ the phase speed. The frequency is given by $\sigma = kc$. For the case of a free upper boundary, internal waves are defined as those with at least one nodal plane, that is, $w = 0$ at least once in $(0, d)$.

It is well known that if $\left(\bar{\varrho} + \frac{g}{\sigma^2}\bar{\varrho}'\right) \geqq 0$ in $(0, d)$, the eigenvalue problem has no internal wave solution. The cutoff frequency

$$\sigma^2 = \max_y \left(\frac{-g\bar{\varrho}'}{\bar{\varrho}}\right) \equiv \sigma_{\mathrm{B.V.}}^2$$

is known as the Brunt-Väisälä frequency.

If we rewrite eqn. (1) in the Liouville normal form by using the new variable

$$w(y) = \left[\frac{\bar{\varrho}(0)}{\bar{\varrho}(y)}\right]^{1/2} f(y)\,, \quad \bar{\varrho}(y) \neq 0 \quad \text{in} \quad (0, d)\,,$$

we have

$$f'' - \left[\left(k^2 + \frac{g}{c^2}\frac{\bar\varrho'}{\bar\varrho}\right) + \frac{\bar\varrho''}{2\bar\varrho} - \frac{\bar\varrho'^2}{4\bar\varrho^2}\right] f = 0 \qquad (4)$$

with either

$$f(0) = f(d) = 0 \qquad (5)$$

or

$$f(0) = 0\,, \quad f'(d) - \left(\frac{1}{2}\frac{\bar\varrho'(d)}{\bar\varrho(d)} + \frac{g}{c^2}\right) f(d) = 0\,. \qquad (6)$$

Clearly, if the coefficient of $f(y)$ in eqn. (4) is non-negative for $y$ in $(0, d)$, the curvature of $f$ will have the same sign as $f$ and there is no internal wave solution to (4), (5) or (4), (6). Thus, there are no internal waves whose frequencies satisfy the inequality

$$\sigma^2 \geqq \max_y \left\{-\frac{g\bar\varrho'}{\bar\varrho} - \left(\frac{\bar\varrho''}{2\bar\varrho} - \frac{\bar\varrho'^2}{4\bar\varrho^2}\right) c^2\right\} \equiv \sigma_0^2\,, \quad y \ \text{ in } \ (0, d)\,.$$

If $(2\bar\varrho\bar\varrho'' - \bar\varrho'^2) \leqq 0$, then $\sigma_0 \geqq \sigma_{\mathrm{B.V.}}$ and no new information results. However, if $(2\bar\varrho\bar\varrho'' - \bar\varrho'^2) > 0$, we have $\sigma_0 < \sigma_{\mathrm{B.V.}}$, so that a lower cutoff frequency is obtained. Further, it also follows from the fact that $\sigma_0^2$ must be non-negative that $c^2$ is bounded from above by

$$c^2 \leqq \max_y \left(\frac{4g\bar\varrho\bar\varrho'}{\bar\varrho'^2 - 2\bar\varrho\bar\varrho''}\right) \equiv c_0^2\,.$$

As an illustration, we compare $\sigma_{\mathrm{B.V.}}$ and $\sigma_0$ for $\bar\varrho(y) = \bar\varrho(0)\,\mathrm{e}^{-\beta y}$, $\beta > 0$. We have

$$\sigma^2_{\mathrm{B.V.}} = g\beta\,; \quad \sigma_0^2 = g\beta - \frac{\beta^2 c^2}{4}\,, \quad c_0^2 = \frac{4g}{\beta}\,.$$

To extend the above result to the case for a compressible fluid, the governing equation is obtained by replacing $k^2$ in eqn. (1) by $k_i^2$, where

$$k_i^2 = k^2 + \frac{c^2}{c_s^2}\left(\frac{g^2}{c^4} - k^2\right)$$

where $k$ is the wave number as defined before, and $c_s$ is the sound speed. Again, by using the normal form for the governing equation, we obtain the following condition for the non-existence of internal waves. For $c < c_s$, there are no internal waves if

$$\sigma^2 \geqq \max_y \frac{c_s^2}{c_s^2 - c^2}\left\{-\frac{g\bar\varrho'}{\bar\varrho} - \frac{c^2}{2}\left(\frac{\bar\varrho''}{\bar\varrho} - \frac{\bar\varrho'^2}{2\bar\varrho^2}\right) - \frac{g^2}{c_s^2}\right\} \equiv \sigma_L^2\,,$$

and for $c > c_s$, there are no internal waves if

$$\sigma^2 \leqq \min_y \frac{c_s^2}{c^2 - c_s^2}\left\{\frac{g\bar\varrho'}{\bar\varrho} + \frac{c^2}{2}\left(\frac{\bar\varrho''}{\bar\varrho} - \frac{\bar\varrho'^2}{2\bar\varrho^2}\right) + \frac{g^2}{c_s^2}\right\} \equiv \sigma_U^2\,.$$

In both cases, the fact that $\sigma^2$ must be positive also induces a bound on $c^2$. We again obtain the cutoff frequencies for $\bar\varrho = \bar\varrho(0)\,\mathrm{e}^{-\beta y}$.

For $c < c_s$,

$$\sigma_L^2 = \frac{c_s^2}{c_s^2 - c^2}\left\{g\beta - \frac{g^2}{c_s^2} - \frac{c^2\beta^2}{4}\right\}$$

and

$$c^2 \leqq \frac{4g}{\beta^2}\left(\beta - \frac{g}{c_s^2}\right) \equiv c_0^2\,, \quad \beta > \frac{g}{c_s^2}\,.$$

If $\beta \leqq \frac{g}{c_s^2}$ there are no internal waves with $c < c_s$.

For $c > c_s$,

$$\sigma_U^2 = \frac{c_s^2}{c^2 - c_s^2}\left\{\frac{c^2\beta^2}{4} + \frac{g^2}{c_s^2} - g\beta\right\},$$

and

$$c^2 \geqq \frac{4g}{\beta^2}\left(\beta - \frac{g}{c_s^2}\right).$$

Acknowledgement: Research supported by the National Research Council of Canada.

### Reference

YIH, C. S., Dynamics of Nonhomogeneous Fluids, Mac-Millan, New York (1965).

Eingereicht am 14. 12. 1978

*Anschrift*: Dr. K. K. TAM, McGill University, Department of Mathematics, Burnside Hall, Montreal, Quebec, H3A 2K6, Canada

ZAMM 60, 57 – 59 (1980)

K. YONEYAMA / T. YUASA / T. NISHIDA

## A Queueing System with c-Random Additional Service Channels

### Introduction

A large number of queueing problems with additional service channels have been solved by various authors. In particular, MURARI (1971) obtains the steady state probabilities of queueing problems with variable number of service channels assuming that when the queue length increases to some undesirable number $m_1$, then another channel is called for its help, and if in spite of two service channels operating the queue length increases to some undesirable number $m_2$ ($> m_1$), then the third channel is called for their help and so on, and in all cases when there is a demand of an additional service channel it is made available instantaneously and is cancelled at the termination of service if there are no units waiting, with the exception of one channel which remains open at all times. More recently SINGH (1978) studies the steady state behaviour of the $M/M/1(M)$ system assuming that when the queue length reaches $M$, then a search for an additional service facility is started and the availability time is exponentially distributed.

In this paper, we modify the results due to MURARI as follows. When the queue length increases to some undesirable number $m_1$, then a search for another channel is started and the availability time of another channel is exponentially distributed. The search continues till the queue length reduces to $m_0 (< m_1)$. If the search is fulfilled, another channel is available and it takes in the head of the queue for service instantaneously. If in spite of two service channels operating the queue length increases to some undesirable number $m_2 (> m_1)$ then a search for the third channel is started and so on.

Thus, this paper studies the *steady state behaviour of a queueing system* under the following assumptions:

I. Customers arrive at a service facility in a POISSON stream with mean rate $\lambda$ and form a queue.

II. The queue discipline is first-come-first-served.

III. The service facility consists of one regular service channel and $c$ additional service channels (a.s.c.'s). The regular service channel is always at the disposal of customers, irrespective of the queue length. However, the number of operating a.s.c.'s at any instant is dependent of the queue length:

(i) If the queue length is zero and the regular service channel is idle, it takes in a unit for service instantaneously on arrival.

(ii) If the queue length is $m_r (1 \leqq r \leqq c)$, and $0 \leqq m_1 \leqq m_2 \leqq \ldots \leqq m_c$) and the number of operating a.s.c.'s is $r - 1$, then a search for the next a.s.c. is started and the availability time (the time when the next a.s.c. is made available, measured from the instant that a search is started) is exponentially distributed with parameter $V$. The search continues till the queue length reduces to $m_{r-1}$ $(1 \leqq r \leqq c)$. If during the small interval $[t, t + \Delta t)$ the search is fulfilled, that is, the next a.s.c. is available, the head of the queue gets the service from the next a.s.c.

(iii) If the queue length is zero at the completion of service at a service channel (regular or additional) then one of the operating a.s.c.'s, if there is any, stops operating.

IV. All the service channels have the same exponential service time distribution with mean service time $\mu^{-1}$.

V. If at any instant the queue length is $m_r$ or $K$ ($> m_c$) and the number of operating a.s.c.'s is $r - 1$ or equal to $c$, respectively, then the arriving customers will be considered lost for the system.

VI. The stochastic processes involved are independent of each other.

Define:

$\mathsf{Q} \equiv$ the steady state probability that the queue length is zero and no service channel is operating;

$\mathsf{P}_{r,n}^{(A)} \equiv$ the steady state probability that the queue length is $n$, the regular service channel is busy, the number of operating a.s.c.'s is $r$ $(0 \leqq r \leqq c-1)$ and a search for the next a.s.c. is in progress;

$\mathsf{P}_{r,n}^{(B)} \equiv$ the steady state probability that the queue length is $n$, the regular service channel is busy, the number of operating a.s.c.'s is $r$ $(0 \leqq r \leqq c)$ and no search for the next a.s.c. is in progress.

Thus by the assumptions imposed on the system

$$\mathsf{P}_{r,n}^{(A)} = 0, \qquad 0 \leqq n \leqq m_r, \qquad 0 \leqq r \leqq c-1,$$

$$\mathsf{P}_{c,n}^{(A)} = 0, \qquad 0 \leqq n \leqq K,$$

$$\mathsf{P}_{r,m_{r+1}}^{(B)} = 0, \qquad 0 \leqq r \leqq c-1.$$

### Steady state solution

The *steady state equations* may be constructed in the usual manner. For $0 \leqq r \leqq c-1$, we have

$$[\lambda + (r+1)\mu + V]\,\mathsf{P}_{r,1+m_r}^{(A)} = (r+1)\mu\mathsf{P}_{r,2+m_r}^{(A)},$$

$$[\lambda + (r+1)\mu + V]\,\mathsf{P}_{r,n}^{(A)} = \lambda\mathsf{P}_{r,n-1}^{(A)} + (r+1)\mu\mathsf{P}_{r,n+1}^{(A)},$$

$$2 + m_r \leqq n \leqq -1 + m_{r+1},$$

$$[(r+1)\mu + V]\,\mathsf{P}_{r,m_{r+1}}^{(A)} = \lambda\mathsf{P}_{r,-1+m_{r+1}}^{(A)} + \lambda\mathsf{P}_{r,-1+m_{r+1}}^{(B)},$$

and

$$\lambda\mathsf{Q} = \mu\mathsf{P}_{0,0}^{(B)},$$

$$(\lambda + \mu)\,\mathsf{P}_{0,0}^{(B)} = \lambda\mathsf{Q} + \mu\mathsf{P}_{0,1}^{(B)} + 2\mu\mathsf{P}_{1,0}^{(B)}, \tag{1}$$

$$(\lambda + \mu)\,\mathsf{P}_{0,n}^{(B)} = \lambda\mathsf{P}_{0,n-1}^{(B)} + \mu\mathsf{P}_{0,n+1}^{(B)}, \quad 1 \leqq n \leqq -2 + m_1,$$

$$n \neq m_0,$$

$$(\lambda + \mu)\,\mathsf{P}_{0,m_0}^{(B)} = \lambda\mathsf{P}_{0,-1+m_0}^{(B)} + \mu\mathsf{P}_{0,1+m_0}^{(B)} + \mu\mathsf{P}_{0,1+m_0}^{(A)},$$

$$(\lambda + \mu)\,\mathsf{P}_{0,-1+m_1}^{(B)} = \lambda\mathsf{P}_{0,-2+m_1}^{(B)},$$

and for $1 \leqq r \leqq c-1$,

$$[\lambda + (r+1)\mu]\,\mathsf{P}_{r,0}^{(B)} = (r+1)\mu\mathsf{P}_{r,1}^{(B)} + (r+2)\mu\mathsf{P}_{r+1,0}^{(B)},$$

$$[\lambda + (r+1)\mu]\,\mathsf{P}_{r,n}^{(B)} = \lambda\mathsf{P}_{r,n-1}^{(B)} + (r+1)\mu\mathsf{P}_{r,n+1}^{(B)},$$

$$1 \leqq n \leqq -1 + m_{r-1}, \qquad 1 + m_r \leqq n \leqq -2 + m_{r+1},$$

$$[\lambda + (r+1)\mu]\,\mathsf{P}_{r,n}^{(B)} = \lambda\mathsf{P}_{r,n-1}^{(B)} + (r+1)\mu\mathsf{P}_{r,n+1}^{(B)} + V\mathsf{P}_{r-1,n+1}^{(A)},$$

$$m_{r-1} \leqq n \leqq -1 + m_r,$$

$$[\lambda + (r+1)\mu]\,\mathsf{P}_{r,m_r}^{(B)} = \lambda\mathsf{P}_{r,-1+m_r}^{(B)} + (r+1)\mu\mathsf{P}_{r,1+m_r}^{(B)} + (r+1)\mu\mathsf{P}_{r,1+m_r}^{(A)},$$

$$[\lambda + (r+1)\mu]\,\mathsf{P}_{r,-1+m_{r+1}}^{(B)} = \lambda\mathsf{P}_{r,-2+m_{r+1}}^{(B)},$$

and

$$[\lambda + (c+1)\mu]\,\mathsf{P}_{c,0}^{(B)} = (c+1)\mu\mathsf{P}_{c,1}^{(B)},$$

$$[\lambda + (c+1)\mu]\,\mathsf{P}_{c,n}^{(B)} = \lambda\mathsf{P}_{c,n-1}^{(B)} + (c+1)\mu\mathsf{P}_{c,n+1}^{(B)},$$

$$1 \leqq n \leqq -1 + m_{c-1}, \qquad m_c \leqq n \leqq K-1,$$

$$[\lambda + (c+1)\mu]\,\mathsf{P}_{c,n}^{(B)} = \lambda\mathsf{P}_{c,n-1}^{(B)} + (c+1) \times$$

$$\times\, \mu\mathsf{P}_{c,n+1}^{(B)} + V\mathsf{P}_{c-1,n+1}^{(A)}, \quad m_{c-1} \leqq n \leqq -1 + m_c,$$

$$(c+1)\mu\mathsf{P}_{c,K}^{(B)} = \lambda\mathsf{P}_{c,K-1}^{(B)}.$$

Define the *probability generating functions*:

$$\mathsf{P}_r^{(A)}(\alpha) = \sum_{n=1+m_r}^{m_{r+1}} \alpha^n\mathsf{P}_{r,n}^{(A)}, \qquad 0 \leqq r \leqq c-1,$$

$$\mathsf{P}_r^{(B)}(\alpha) = \sum_{n=0}^{-1+m_{r+1}} \alpha^n\mathsf{P}_{r,n}^{(B)}, \qquad 0 \leqq r \leqq c-1, \tag{2}$$

$$\mathsf{P}_c^{(B)}(\alpha) = \sum_{n=0}^{K} \alpha^n\mathsf{P}_{c,n}^{(B)}.$$

Multiplying the equations (1) by appropriate powers of $\alpha$ and using (2), we have

$$\mathsf{P}_r^{(A)}(\alpha) = \frac{\lambda(1-\alpha)\,\alpha^{1+m_{r+1}}\,\mathsf{P}_{r,m_{r+1}}^{(A)} + \lambda\alpha^{1+m_{r+1}}\,\mathsf{P}_{r,-1+m_{r+1}}^{(B)}}{[\lambda + (r+1)\mu + V]\,\alpha - \lambda\alpha^2 - (r+1)\mu} -$$

$$- \frac{(r+1)\,\mu\alpha^{1+m_r}\mathsf{P}_{r,1+m_r}^{(A)}}{[\lambda + (r+1)\mu + V]\,\alpha - \lambda\alpha^2 - (r+1)\mu}, \qquad 0 \leqq r \leqq c-1,$$

$$\mathsf{P}_0^{(B)}(\alpha) =$$

$$= \frac{\lambda\alpha\mathsf{Q} + 2\mu\alpha\mathsf{P}_{1,0}^{(B)} + \mu\alpha^{1+m_0}\mathsf{P}_{0,1+m_0}^{(A)} - \mu\mathsf{P}_{0,0}^{(B)} - \lambda\alpha^{1+m_1}\mathsf{P}_{0,-1+m_1}^{(B)}}{(\lambda+\mu)\,\alpha - \lambda\alpha^2 - \mu},$$

$$\mathsf{P}_r^{(B)}(\alpha) =$$

$$\frac{\begin{pmatrix}(r+2)\,\mu\alpha\mathsf{P}_{r+1,0}^{(B)} + (r+1)\,\mu\alpha^{1+m_r}\mathsf{P}_{r,1+m_r}^{(A)} + V\mathsf{P}_{r-1}^{(A)}(\alpha)\\ -(r+1)\,\mu\mathsf{P}_{r,0}^{(B)} - \lambda\alpha^{1+m_{r+1}}\mathsf{P}_{r,-1+m_{r+1}}^{(B)}\end{pmatrix}}{[\lambda + (r+1)\mu]\,\alpha - \lambda\alpha^2 - (r+1)\mu},$$

$$1 \leqq r \leqq c-1, \tag{3}$$

$$\mathsf{P}_c^{(B)}(\alpha) = \frac{\lambda(1-\alpha)\,\alpha^{K+1}\mathsf{P}_{c,K}^{(B)} + V\mathsf{P}_{c-1}^{(A)}(\alpha) - (c+1)\,\mu\mathsf{P}_{c,0}^{(B)}}{[\lambda + (c+1)\mu]\,\alpha - \lambda\alpha^2 - (c+1)\mu}.$$

Since $\mathsf{P}_r^{(A)}(\alpha)$, $0 \leqq r \leqq c-1$, and $\mathsf{P}_r^{(B)}(\alpha)$, $0 \leqq r \leqq c$ are regular on $\alpha$-plane, the zeros of the numerator and the denominator of the right-hand side of $\mathsf{P}_r^{(A)}(\alpha)$ and $\mathsf{P}_r^{(B)}(\alpha)$, which are given in (3), must coincide with each other, giving $4c + 2$ equations. Solving this set of equations together with the steady state equation

$$\lambda\mathsf{Q} = \mu\mathsf{P}_{0,0}^{(B)},$$

we can determine the $4c + 3$ unknowns occuring on the numerator of the right-hand side of (3) as follows.

For $0 \leqq r \leqq c-1$,

$$\mathsf{P}_{r,1+m_r}^{(A)} = \left[\prod_{i=1}^{r} g_i\right] \mathsf{Q}/(r+1)\,\alpha_0(b_0 + c_0 - b_0d_0 - 1),$$

$$\mathsf{P}_{r,m_{r+1}}^{(A)} = (r+1)\,\alpha_0 a_r \mathsf{P}_{r,1+m_r}^{(A)},$$

$$\mathsf{P}_{r,-1+m_{r+1}}^{(B)} = (r+1)\,\alpha_0 b_r \mathsf{P}_{r,1+m_r}^{(A)}, \tag{4}$$

$$\mathsf{P}_{r+1,0}^{(B)} = (r+1)(b_r - 1)\,\mathsf{P}_{r,1+m_r}^{(A)}/(r+2),$$

and

$$\mathsf{P}_{0,0}^{(B)} = \mathsf{Q}/\alpha_0,$$

$$\mathsf{P}_{c,K}^{(B)} = (c_c - b_cd_c)\left[\prod_{i=1}^{c} g_i\right] \mathsf{Q}/\alpha_c^{K+1}\,(b_0 + c_0 - b_0d_0 - 1),$$

where

$$a_r = \sum_{i=0}^{-1-m_r+m_{r+1}} (\beta_{r+1}/\gamma_{r+1})^i/\gamma_{r+1}\beta_{r+1}^{-m_r+m_{r+1}}, \quad 0 \leqq r \leqq c-1,$$

$$b_r = \left[\gamma_{r+1} \sum_{i=0}^{-m_r+m_{r+1}} (\beta_{r+1}/\gamma_{r+1})^i - \sum_{i=0}^{-1-m_r+m_{r+1}} (\beta_{r+1}/\gamma_{r+1})^i\right]$$

$$/\gamma_{r+1}\beta_{r+1}^{-m_r+m_{r+1}}, \qquad 0 \leqq r \leqq c-1,$$

$$c_0 = 1 - \sum_{i=1}^{m_0} \alpha_0^i, \qquad d_0 = 1 - \sum_{i=1}^{m_1} \alpha_0^i,$$

$$c_r = 1 - \sum_{i=0}^{m_r} \alpha_r^i, \quad 1 \leqq r \leqq c-1, \qquad d_r = 1 - \sum_{i=0}^{m_{r+1}} \alpha_r^i,$$

$$1 \leqq r \leqq c-1,$$

$$e_{r-1} = -(V/\lambda) \sum_{i=1}^{m_{r-1}} \alpha_r^i - \alpha_r + \beta_r\gamma_r, \qquad 1 \leqq r \leqq c,$$

$$f_{r-1} = -(V/\lambda) \sum_{i=1}^{m_r} \alpha_r^i - \alpha_r + \beta_r \gamma_r , \qquad 1 \leqq r \leqq c ,$$

$$g_r = [(V/\lambda)\, a_{r-1} \alpha_r^{1+m_r r} + b_{r-1} f_{r-1} - e_{r-1}]/(\alpha_r - \beta_r) \times$$
$$\times (\gamma_r - \alpha_r)(b_r d_r - c_r) , \qquad 1 \leqq r \leqq c ,$$

$$\alpha_r = (r+1)\, \mu/\lambda , \qquad 0 \leqq r \leqq c ,$$

$$\beta_r = \{\lambda + r\mu + V + [(\lambda + r\mu + V)^2 - 4r\lambda\mu]^{1/2}\}/2\lambda ,$$
$$1 \leqq r \leqq c ,$$

$$\gamma_r = \{\lambda + r\mu + V - [(\lambda + r\mu + V)^2 - 4r\lambda\mu]^{1/2}\}/2\lambda , \ 1 \leqq r \leqq c ,$$

and $\mathsf{Q}$ can be determined by the normalization condition

$$\mathsf{Q} + \sum_{r=0}^{c-1} \mathsf{P}_r^{(A)}(1) + \sum_{r=0}^{c} \mathsf{P}_r^{(B)}(1) = 1 .$$

Comparing the coefficient of $\alpha^n$ on both sides of (3) and using (4), we have, for $c \geqq 1$,

$$\mathsf{P}_{r,n}^{(A)} = \left[ \sum_{i=0}^{n-1-m_r} (\beta_{r+1}/\gamma_{r+1})^i \right] \mathsf{P}_{r,1+m_r}^{(A)} / \beta_{r+1}^{n-1-m_r} , \qquad 0 \leqq r \leqq c - 1 ,$$
$$2 + m_r \leqq n \leqq m_{r+1} ,$$

$$\mathsf{P}_{0,n}^{(B)} = b_0 \left[ \sum_{i=1}^{-n+m_1} \alpha_0^i - \sum_{i=1}^{-n+m_0} \alpha_0^i \right] \mathsf{P}_{0,1+m_0}^{(A)} , \qquad 1 \leqq n \leqq -1 + m_0 ,$$

$$\mathsf{P}_{0,n}^{(B)} = b_0 \mathsf{P}_{0,1+m_0}^{(A)} \sum_{i=1}^{-n+m_1} \alpha_0^i , \qquad m_0 \leqq n \leqq -1 + m_1 ,$$

$$\mathsf{P}_{c,n}^{(B)} = \alpha_c^{K-n} \mathsf{P}_{c,K}^{(B)} , \qquad 1 + m_c \leqq n \leqq K ,$$

and especially, in the case $c = 1$ and $m_0 = 0$,

$$\mathsf{P}_{0,1}^{(A)} = \mathsf{Q}/\alpha(b_0 - b_0 d_0 - 1) ,$$

$$\mathsf{P}_{1,n}^{(B)} = \mathsf{P}_{1,0}^{(B)}/\gamma_1^n + \sum_{i=1}^{n} \left[ \sum_{j=K+i-n-1}^{K} h_{n-i+1}^{(j)} \right] \Big/ \gamma_1^i , \qquad 1 \leqq n \leqq m_1,$$

where

$$h_n^{(i)} = \left[ -(V/\lambda)\, b_0 \sum_{j=l}^{n+i-K} \beta^{-j} + (1 - b_0)(1 - \gamma_1)\, \beta_1^{K-n-i} \right] \times$$
$$\times \alpha_0 \alpha_1^{i-K-1} \mathsf{P}_{0,1}^{(A)} ,$$

$$h_n^{(K-n)} = (b_0 - 1)\, \gamma_1 \alpha_1^{-n-1} .$$

### Particular case

The results due to Murari (1971) can be obtained by exchanging $\mathsf{P}_{r,n}^{(B)}$ $(0 \leqq r \leqq c-1,\ 0 \leqq n \leqq -1 + m_{r+1};\ r = c,\ 0 \leqq n \leqq K)$ and $-1 + m_{r+1}$ $(0 \leqq r \leqq c - 1)$ with $\mathsf{P}_{r,n}$ and $m_{r+1}$, respectively, and letting $V$ tend to infinity in our model.

### References


1 Murari, K., Some c-additional service channels, limited space queueing problem, ZAMM 51, 315–318 (1971).

2 Singh, K. Bidhi, A queueing system with random additional service facility, R.A.I.R.O. 12, 311–318 (1978).



*Anschrift*: Dr. Kanji Yoneyama; Dr. Teru Yuasa; Dr. Toshio Nishida, Osaka University, Faculty of Engineering, Department of Applied Physics Osaka 565, Japan


# BUCHBESPRECHUNGEN

**Strehmel, K.,** Näherungsmethoden zur Ermittlung von Transsonischen Profilströmungen mit Verdichtungsstößen. Abhandlungen der Sächsischen Akademie der Wissenschaften zu Leipzig, Band 53, Heft 4. Berlin, Akademie-Verlag 1978. 78 S., M 23,–.

In der vorliegenden Abhandlung wird die Hodographenmethode benutzt, um bei Unterschallanströmung stoßbehaftete transsonische Profilumströmungen zu berechnen. Wesentlich ist, daß hierbei im Überschallteil keine Grenzlinien auftreten. Besondere Beachtung finden der Stoßbeginn im Überschallfeld sowie das Auftreten des Stoßes am stetig gekrümmten Profil. Der Verfasser legt großen Wert auf die Herleitung numerisch übersichtlicher Ausdrücke für seine Lösung. Die Arbeit stellt einen wesentlichen Fortschritt auf diesem schwierigen Gebiet dar, und man kann insbesondere auf weitere Beispiele, die mit dieser Methode gerechnet werden, gespannt sein.

Karlsruhe J. Zierep

Finite Elements in Fluids. Vol. 3. Edited by **Gallagher, R. H./Zienkiewicz, O. C./Oden, J. T./Morandi Cecchi, M./Taylor, C.,** Chichester-New York-Brisbane-Toronto. John Wiley & Sons 1978. 396 S., £ 17.50.

Dieser Sammelband enthält 21 überarbeitete Artikel der 2. Konferenz „Finite Elemente für Strömungsprobleme", die 1976 auf St. Margherita Ligure in Italien stattfand. Er zeigt die breite Anwendbarkeit und Vorteile der Methode der Finiten Elemente (FEM) in Strömungsproblemen. Dargestellt werden u. a. Anwendungen auf kompressible und inkompressible Strömungen, Strömungen durch poröse Medien, Diffusions-Konvektionsprobleme, Wärmekonvektion, Umströmung von Körpern im schallgeschwindigkeitsnahen Bereich, freie Randwertaufgaben, Plasmaströmungen. Dabei überwiegen Beiträge zu stationären, linearisierten Problemen. Im Vordergrund stehen praktische Gesichtspunkte der FEM (Berücksichtigung geometrischer Verhältnisse, numerische Realisierungen), numerische Experimente und z. T. Modellierungsaspekte. Einige Arbeiten enthalten auch theoretische Untersuchungen zur Konvergenz und zum Fehlerverhalten der FEM.

Weniger dargestellt sind instationäre Probleme. In diesem Zusammenhang ist der Artikel von Gresho, Lee, Sani von Interesse, in dem an mehreren Beispielen die negative Auswirkung des „mass lumping" (Diagonalisierung der Matrix bei der Zeitableitung) auf die Genauigkeit der Ergebnisse gezeigt wird. Eine praktikable Stabilitätstheorie fehlt hier aber.

Hervorzuheben sind weiterhin mehrere Artikel zu transonischen Umströmungen (Glowinski, Periaux, Prionneau; Shen; Hafez, Wellford, Murman; Carey), wo man versucht, mit der FEM dem Typwechsel der Gleichungen und den Stoßwellen gerecht zu werden, sowie ein Beitrag zum Plasmagleichgewicht im Tokamak (Akin, Wooten), in dem die Möglichkeiten der FEM zur Berücksichtigung der Geometrie des Plasmas dargestellt werden.

Beim Vergleich mit Differenzenmethoden werden meist nur die einfachsten von ihnen herangezogen (sogar ein instabiles: im Artikel von Gresho, Lee, Sani). Arbeiten der Schulen um Samarsky, Godunow oder Janenko zu Differenzenverfahren, in denen Differenzenschemata mit variablen Gitternetzen – angepaßt an Geometrie oder Strömung – konstruiert, untersucht und angewendet werden, scheinen unbekannt zu sein.

Der Band ist hervorragend ausgestattet mit einer Vielzahl von Grafiken zur Illustration der Rechenergebnisse. Er wird abgeschlossen durch einen Übersichtsartikel (Norrie, de Vries) mit 218 Literaturstellen (davon 3 nicht englisch), der eine detaillierte Übersicht über Entwicklungen und Anwendungen der FEM in Strömungsproblemen gibt.

Berlin R. Kluge, G. Stoyan

**Cedergren, H. R.,** Seepage, Drainage, & Flow Nets. 2nd. edition. New York-London-Sydney-Toronto, John Wiley & Sons 1977. XIII, 534 S., £ 23.15.

In der vorliegenden 2. Auflage des 1967 erschienenen Werkes werden Methoden der Analyse und des Entwurfs von Einrichtungen zur Beherrschung von Sickerwasser- und Grundwasserströmungen für eine Vielzahl von Anwendungsgebieten wie z. B. Böschungsstabilisierung, Erddämme, Deiche, Wehre, Staumauern, Halden, Kohlepfeiler, Straßen, Flugpisten, Gründungen, Entwässerungspfähle, Baugruben, Drainagen in der

Landwirtschaft, Infiltrationseinrichtungen und Grundwasserabsenkung durch Brunnen vorgelegt. Dabei werden Methoden benutzt, die keine hohen theoretischen und mathematischen Ansprüche an den Leser stellen.

Im ersten Teil des Buches werden die Theorie der Sickerströmung und die Ermittlung der das Sickerwasserströmungsfeld charakterisierenden Kennwerte, z. B. Durchlässigkeitsmerkmale, und Randbedingungen, wie Anisotropien und Inhomogenitäten, mit der Konstruktion der Sickerwasserströmungsnetze dargestellt, wobei auch der Einfluß von turbulenter Strömung auf die Ausbildung des Strömungsnetzes Berücksichtigung gefunden hat. Der zweite Zeil zeigt die praktische Bedeutung von Sickerwasserströmungen für den Entwurf von Filter und Drainagen sowie von Meßeinrichtungen, die für die Kontrolle des Sickerwasserströmungsfeldes geeignet sind. Hinweise auf Möglichkeiten der Sanierung von Schäden an Bauwerken, die durch Sickerwasserströmung entstanden sind, auf die Vergrößerung der seismischen Standsicherheit von Staudämmen durch Anordnung von entsprechenden Drainagen, das Vorlegen von Tabellen und Diagrammen zur Analyse von Sickerwasserströmungen und zum Entwurf von Drainagen regen den Leser an, die außerordentliche Bedeutung der Sickerwasserströmung für viele Standsicherheitsprobleme zu sehen und die Prinzipien der Sickerwasserströmung auf die vielschichtigen Probleme anzuwenden.

Der mathematischen Modellierung der Strömungsvorgänge mit ihren analytischen, modelltechnischen und numerischen Lösungsmöglichkeiten (Analogieverfahren, Digitalrechner, Finite-Element-Methode, Verwendung von EDV-Programmen) wird in nur sehr geringem Umfang Platz eingeräumt. Dem Autor gelingt es, die Praktiker anzusprechen und ihnen in umfangreicher Weise ausführliche Beispiele für Sickerwasserströmungsbeeinflussung von Bauwerken und für die Abwehr ihrer Schädigung durch Strömungsvorgänge in die Hand zu geben.

Freiberg H.-J. Schaef

**Lai, W. M./Rubin, D./Krempl, E.,** Introduction to Continuum Mechanics. Revised edition in SI/Metric units. Pergamon Unified Engineering Series 17. Oxford-New York-Toronto-Sydney-Paris-Frankfurt. Pergamon Press 1978. XIV, 310 S., \$ 30.00. ℱ 15.00 (Hardcover). \$ 15.00. £ 7.50. (Flexicover)

Die vorgelegte Einführung in die Kontinuumsmechanik enthält eine ausführliche Darlegung der Tensorrechnung für Tensoren im 3dimensionalen euklidischen Raum. Es schließen sich die Kinematik des Kontinuums, in der Hauptsache spezialisiert auf kleine Verzerrungen, und die Theorie des Spannungstensors an. Die konstitutiven Gleichungen werden für 3 Modelle behandelt: den linear elastischen festen Körper, linear viskose Flüssigkeiten und sogenannte inkompressible einfache (nichtnewtonsche) Flüssigkeiten. Die Lösungen für einige einfache Randwertprobleme werden gezeigt. Das Buch zeichnet sich durch klare Begriffsbestimmungen und Ausdrucksweise aus. So ist z. B. die Gegenüberstellung und stets klare Unterscheidung zwischen Lagrangescher und Eulerscher Darstellung wichtig für das Verständnis durch den Anfänger. Nützlich sind auch die vielen Beispiele und Übungsaufgaben. Das Buch ist gut geeignet, dem Anfänger eine solide Grundlage für die Einarbeitung in die Methoden der Ingenieurpraxis zu vermitteln.

Magdeburg U. Fischer

**Madelung, O.,** Introduction to Solid-State Theory. Springer Series in Solid-State Sciences 2. — Übersetzung aus dem Deutschen. Berlin-Heidelberg-New York, Springer-Verlag 1978. XI, 486 S., 144 Abb., DM 59,—. US \$ 29.50.

Dieses von B. C. Taylor ins Englische übersetzte Werk ist aus der „Festkörpertheorie I bis III" des Autors (Heidelberger Taschenbücher 104, 109, 126 bei Springer 1972, 1973) bei teilweiser Neubearbeitung hervorgegangen. Dabei wurden die früheren Anhänge über Gruppentheorie und Greensche Funktionen weggelassen. Der einheitliche Ausgangspunkt bei der Entwicklung der Grundlagen dieses umfangreichen theoretischen Stoffgebietes, das Physikern, Materialwissenschaftlern und Elektroingenieuren angeboten wird, ist die Beschreibung mittels delokalisierter und lokalisierter Zustände sowie Elementaranregungen. Diese Methode gewährleistet einen systematischen Zugang, wodurch das Fachgebiet durchsichtig und übersichtlich gestaltet werden kann. Außerdem wird dadurch eine anschaulich-bildliche Formulierung vieler Elementarprozesse erreicht, so daß auch das Verständnis beachtlich erleichtert wird. Die damit eingehandelten Nachteile hinsichtlich des gesteckten Rahmens für die Anwendung nimmt der Autor dafür gerne in Kauf, zumal eine umfangreiche, aktuelle Bibliographie auf aus diesem Rahmen herausfallende Problemkreise hinweist.

Die wichtigsten Stoffgebiete sind: Ein-Elektron-Approximation, Elementaranregungen, Elektron-Phonon-Wechselwirkungen (Transportphänomene), Elektron-Elektron-Wechselwirkungen bei Austausch virtueller Phononen (Supraleitung), Elektron-Photon-Wechselwirkungen, Phonon-Photon-Wechselwirkungen, Phonon-Phonon-Wechselwirkungen, lokalisierte Zustände, ungeordnete Gitter etc. Das mit Absicht auf die abstrakten quantentheoretischen Methoden verzichtende Buch ist vor allem denjenigen zu empfehlen, die einen anschaulichen Zugang zur Festkörpertheorie suchen.

Jena E. Schmutzer

**Broek, D.,** Elementary Engineering Fracture Mechanics. Alphen aan den Rijn, Sijthoff & Noordhoff 1978. XIII, 437 S., Dfl 120.00, \$ 60.00 (cloth), Dfl 40.00, \$ 20.00 (paper)

Diese zweite Auflage stimmt mit der ersten (vgl. Rezension in ZAMM 58 (1978), S. 423) im wesentlichen überein, enthält aber einige Ergänzungen und Aktualisierungen. Geblieben ist ein ausgezeichnetes Buch von einführendem Charakter, das eine allgemeine Übersicht über das Gebiet vermittelt, vom phänomenologischen Befund bis zur physikalischen Interpretation, immer mit dem Blick auf technische Anwendungen und deren Hintergrund.

Berlin W. Heinrich

**Klotter, K.,** Technische Schwingungslehre. Band 1: Einfache Schwinger. Teil A: Lineare Schwingungen. 3., neubearb. u. erweiterte Auflage, herausgeg. mit Unterstützung durch G. Benz. Berlin-Heidelberg-New York, Springer-Verlag 1978. XIII, 425 S., DM 64,—; US\$ 32,—

Die „Technische Schwingungslehre" Karl Klotters, deren erster Band 1938 und in zweiter Auflage 1951 erschien, ist zu einem festen Begriff für alle geworden, die mit Schwingungserscheinungen in der Technik zu tun haben. „Der Klotter" wurde nicht nur von Studenten, sondern auch von in Forschung und Praxis an Schwingungsproblemen Arbeitenden immer wieder zur Hand genommen. Zu der längst überfälligen Neuauflage schreibt der Autor: „Die zweite Auflage einfach unverändert zu reproduzieren, ging nicht an: Eine Schwingungslehre auch nur des „einfachen Schwingers" darf sich nicht mehr mit den linearen Vorgängen begnügen; die nichtlinearen bilden nun schon seit vielen Jahren einen wesentlichen Bestandteil des Stoffes, den der Benutzer braucht und sucht." So mußte der erste Band in zwei Teilbände aufgeteilt werden, deren erster, vom Umfang des gesamten früheren ersten Bandes, jetzt vorliegt und den linearen einfachen Schwinger behandelt. Der Autor hat ihn, unterstützt von G. Benz, völlig umgearbeitet, wobei seine nun etwa fünfzigjährige Lehrerfahrung reiche Früchte trägt.

Noch stärkeres Gewicht als in der ersten Auflage hat das systematische Aufstellen der Bewegungsgleichungen mechanischer Schwingungen auf der Grundlage der Prinzipe der Mechanik erhalten. Laufend findet ein Bezug auf die derzeit gültigen DIN-Normen statt. Den Hauptteil des Buches bilden die verschiedenen Arten von nichtautonomen oder fremderregten Schwingungen, bei denen die Zeit explizit in den Bewegungsgleichungen auftritt. Dabei werden als stör-erregte oder erzwungene Schwingungen solche bezeichnet, bei denen die Störfunktionen additiv in den Bewegungsgleichungen auftreten (in begründetem Unterschied zu der im Normblatt DIN 1311 vorgeschlagenen Bezeichnung Quellerregung). Schwingungen, bei denen die Störfunktionen als Koeffizienten bei der Schwingungsfunktion oder ihren Ableitungen auftreten und an deren ersten Untersuchungen der Autor vor etwa 40 Jahren beteiligt war, werden dem neueren Sprachgebrauch entsprechend als parametererregte, gelegentlich auch als rheolineare oder rheonichtlineare Schwingungen bezeichnet.

Die Darstellung der parametererregten Schwingungen und der Fragen des An- und Auslaufens und des Resonanzdurchgangs wurden wesentlich erweitert. Völlig neu aufgenommen wurde eine 70seitige Behandlung der Schwingungen unter stoßartigen Einwirkungen oder Schockeinwirkungen, die bisher kaum in Lehrbücher der Schwingungstheorie Eingang gefunden und hier eine erstmalige systematische auf Arbeiten von K.-E. MEIER-DÖRNBERG basierende Darstellung erfahren haben. Im Mittelpunkt stehen dabei die Gegenüberstellung von Operationen im Originalraum für Zeitfunktionen und von Operationen im Bildraum für Spektralfunktionen und die Klassifikation der Schockeinwirkungen nach den Maximalwerten der einzelnen Ableitungen. Auch an manchen anderen Stellen, etwa bei der Untersuchung der Lösungen inhomogener Differentialgleichungen mit periodischen Koeffizienten, sind unveröffentlichte Ergebnisse eingeflossen.

Ein Charakteristikum der Neuauflage ist die stärkere Einbeziehung numerischer Lösungsverfahren, wobei auf die unveränderte Wichtigkeit analytischer Näherungsverfahren hingewiesen wird. Die Neuauflage dürfte für lange Zeit ein grundlegendes Werk der Schwingungslehre darstellen.

Berlin G. SCHMIDT

**Thirring, W.,** Classical Dynamical Systems. A Course in Mathematical Physics 1. Wien-New York, Springer-Verlag 1978. XII, 258 S., 58 Abb., S 304,–. DM 44,–.

Das vorliegende Lehrbuch, das erste eines vierbändigen Kurses der Mathematischen Physik, wendet sich an Physiker, die tiefer in theoretische Schlußweisen eindringen möchten. Es ist aber auch Mathematikern zu empfehlen, die sich um theoretisch-physikalische Kenntnisse bemühen, dabei aber mathematisch sicheren Boden nur ungern verlassen möchten. Die Anforderungen, die an den Leser gestellt werden, sind nicht gering. Das Buch beginnt nach einer kurzen, aber überdenkenswerten Einleitung mit Grundlagen der Differentialgeometrie (Mannigfaltigkeiten, Faserbündel, Differentialformen, Integration), um danach die Mechanik mit dem Studium der kanonischen Transformationen sowie der HAMILTONschen Gleichungen einzuleiten. Besonders zu vermerken ist die sorgfältige Behandlung des klassischen Streuproblems und der Störungstheorie. Die beiden folgenden Kapitel sind der Integration von Bewegungsgleichungen gewidmet. Das erste konzentriert sich auf den gegenwärtigen Stand des KEPLER-NEWTONschen Mehrkörperproblems, das letztere behandelt eine instruktive Auswahl an relativistischen Bewegungsgleichungen in gegebenen elektromagnetischen bzw. Gravitationsfeldern. Das Buch schließt mit einem kurzen Exkurs in die „Struktur von Raum und Zeit", der wohl nur als ein Zwischenspiel vor dem Band „Klassische Feldtheorie" gedacht worden ist.

Die Fülle der wohldurchdachten Beispiele, Übungen, Problemstellungen und interessanten Bemerkungen ist erstaunlich. Der Autor überzeugt den Leser durch anregende und kluge Argumentation, daß auch die klassischen Theorien in Bewegung sind und keineswegs in erstarrter Abgeschlossenheit ruhen.

Ich glaube, daß WALTER THIRRING, ein erfahrener Hochschullehrer, der selbst Wichtiges zur Bereicherung der Mathematischen und Theoretischen Physik beigetragen hat, mit diesem Buch und dieser Buchserie etwas Wesentliches zum Stil und zur Entwicklung der heutigen Theoretischen Physik gesagt hat.

Leipzig A. UHLMANN

**Pippard, A. B.,** The Physics of Vibration. Vol. 1. Part 1: The Simple Classical Vibrator. Cambridge-London-New York-Melbourne, Cambridge University Press 1978. 431 S., £ 22.50 A H/C.

Dieses Werk des bekannten englischen Physikers PIPPARD, der u. a. durch Arbeiten zur Thermodynamik, zur statistischen Mechanik und zur Festkörpertheorie hervorgetreten ist, stellt ein Lehrbuch auf mittlerem Niveau dar, das in Stoffauswahl, Anlage und Darstellungsstil außergewöhnlich ist. Schwingungen der verschiedensten Arten treten beinahe überall in der Physik auf, und die Formeln etwa des harmonischen Oszillators beherrschen weite Teile der klassischen Physik und beispielsweise der Quantenelektrodynamik. Das Buch versteht sich nach den Intentionen des Verfassers als Brücke zwischen dem elementaren Wissen jedes Physikers und Ingenieurs und dem Spezialwissen der verschiedensten Teildisziplinen, wie es in Monographien auf hohem Niveau abgehandelt wird.

Der vorliegende erste Band umfaßt einfache Systeme (mit nur einer schwingenden Variablen), und zwar klassische — d. h. nichtquantenmechanische — sowie klassisch behandelbare Systeme. Aber bereits damit wird eine riesige Fülle an Stoff erfaßt und an zahlreichen Beispielen erläutert, wie die folgenden Stichwörter zeigen: Harmonische, anharmonische und subharmonische Schwingungen, Relaxationsprozesse, parametrische Anregung, Schwingkreise, negative Widerstände, Musikinstrumente, Kernspinresonanz, Zyklotronresonanz im Festkörper, Heliconen, Neutronenstreuung, JOSEPHSON-Effekt der Supraleitung. Die Wellentheorie bleibt außerhalb des Rahmens des Buches, abgesehen von stehenden Wellen und elektromagnetischen Hohlraumschwingungen.

Das mathematische Niveau ist elementar gehalten; vorausgesetzt werden nur Kenntnisse über Differentialgleichungen. Im Buch werden die einfachsten Grundlagen der Analyse linearer Systeme mittels Funktionentheorie, der Stabilitätstheorie sowie der Theorie des Rauschens bis zur Spektralanalyse stochastischer Funktionen und zum WIENER-KHINTCHIN-Theorem mit behandelt.

Die Darstellung ist breit angelegt und sehr klar, dabei stehen immer die physikalischen Zusammenhänge und weniger die mathematischen Details im Vordergrund. Zahlreiche Abbildungen ergänzen den Text. Ein umfangreiches Verzeichnis weiterführender Literatur und ein Sachregister beschließen den Band, dessen Satz, Druck und äußere Ausstattung vorzüglich sind. Das Werk ist weiten Kreisen von Mathematikern, Physikern, Chemikern und Ingenieuren sehr zu empfehlen; es ist auch zum Selbststudium besonders geeignet. Man darf auf den zweiten Band gespannt sein, der die quantenmechanischen Systeme und Systeme mit vielen Freiheitsgraden behandeln wird.

Dresden G. VOJTA

**Tondl, A.,** On the Interaction between Self-Excited and Parametric Vibrations. Prague 1978. Monographs and Memoranda 25. Special Printing of the National Research Institute for Machine Design, Bechovice. 127 S.

Mit dem vorliegenden Buch setzt der bekannte Prager Schwingungsforscher A. TONDL die Reihe seiner Monografien auf dem Gebiet der nichtlinearen Schwingungen fort. Er behandelt in ihm die interessanten Schwingungserscheinungen, die durch Wechselwirkung von Parameter- und Selbsterregung entstehen und bisher im wesentlichen von ihm selbst gefunden worden sind, wobei ein großer Teil der erhaltenen Ergebnisse noch nicht oder nur in tschechischer Sprache veröffentlicht wurde. Spezielle Beispiele dieser Wechselwirkung sind die parametererregten Querschwingungen längs pulsierend belasteter Stäbe, Masten oder Türme in einem strömenden Medium, etwa bei zusätzlicher Windbelastung, parametererregte Schwingungen von Pendeln in strömendem Medium oder mit trockener Reibung sowie selbst- und parametererregte Schwingungen durch ungleiche Trägheitsmomente bei Turbomaschinen.

Zur Untersuchung werden analytische Näherungsmethoden, vor allem die Methode der harmonischen Balance, sowie vergleichende Analogrechnungen benutzt. Die vom Autor bei erzwungenen Schwingungen erfolgreich angewendeten Schwingungscharakteristiken der Skelettlinie und der Grenzenveloppe werden auch bei den hier behandelten Problemen häufig zugrunde gelegt. Bei Systemen mit einem Freiheitsgrad wird die zusätzliche Wirkung einer trockenen Reibung berücksichtigt, sie reduziert die Amplitude der selbsterregten wie auch der parametererregten Schwingungen. Besonders untersucht werden Systeme mit periodisch veränderlichen Rückstellkraftanteilen (auch mit nichtlinearer Parametererregung bezeichnet). Bei Systemen mit zwei Freiheitsgraden ergeben sich ebenso durch die Möglichkeit von Kombinationsresonanzen als auch durch verschiedene Zustände selbsterregter Schwingungen qualitative Unterschiede. Ein grundsätzliches Ergebnis ist, daß in den Bereichen der Parametererregung erster und zweiter Art die selbsterregten Schwingungen vollständig unterdrückt werden können.

Die in der Monografie veröffentlichten Ergebnisse des Autors stellen einen wesentlichen Beitrag zum Forschungsgebiet der nichtlinearen mechanischen Schwingungen dar. Dem Buch ist weite Verbreitung zu wünschen.

Berlin G. SCHMIDT

**Nöbeling, G.,** Integralsätze der Analysis. de Gruyter Lehrbuch. Berlin-New York, Walter de Gruyter 1979. 117 S., DM 28,–.

In der Regel verbleibt im Hochschulstudium innerhalb der üblichen Kursvorlesungen zur Differential- und Integralrechnung viel zu wenig Zeit für eine sorgfältige und ausführliche Darlegung der Integralsätze von GAUSS und STOKES. Mit diesem Lehrbuch, das sich sowohl für ein Selbststudium als auch für seminaristische Durcharbeitung vorzüglich eignet, läßt sich das Versäumte nachholen. Der Autor bereitet dazu in gepflegter Sorgfalt und Klarheit grundlegende Begriffe der globalen Analysis unter Beschränkung auf $n$-Mannigfaltigkeiten der Klasse $C^2$ im $E^m$ $(m > n)$ auf. Es folgen alternierende Differentialformen $p$-ten Grades und RIEMANNsche Integration auf Mannigfaltigkeiten unter gewissenschafter Analyse der Invarianz- und Transformationseigenschaften dieser Grundbegriffe gegenüber $C^2$-Diffeomorphismen. Auf diesem Fundament wird schließlich der allgemeine STOKESsche Satz in der CARTANschen Fassung entwickelt und in einigen Spezialfällen und Anwendungen zum Einsatz gebracht. Der Anhang enthält Ergänzungen, begriffliche Hilfsmittel und ein Symbolregister. Das Literaturverzeichnis mit 10 Titeln ist etwas knapp bemessen; es fehlen für den in der Forschung tätigen Mathematiker Hinweise auf weiterführende Untersuchungen (wie z. B. von K. KRICKEBERG — Math. Nachr. 10—12) unter schwächeren Voraussetzungen.

Leipzig R. KLÖTZLER

**Meis, Th./Marcowitz, U.**, Numerische Behandlung partieller Differentialgleichungen. Hochschultext. Berlin-Heidelberg-New York, Springer-Verlag 1978. VIII, 452 S., 31 Abb., DM 38,—. US $ 20.90.

Bei der numerischen Lösung von partiellen Differentialgleichungen wird die Aufgabe im allgemeinen durch eine diskrete Problemstellung angenähert. Im Buch werden die Diskretisierungsverfahren für die Anfangswertaufgaben und für die Randwertaufgaben und die in diesem Zusammenhang stehenden Fragen wie Stabilität und Konvergenz besprochen. Es werden die Differenzenverfahren, die Variationsmethoden, die Extrapolationsmethoden, das RITZ-Verfahren und das Kollokationsverfahren behandelt. Die numerische Lösung der durch die Diskretisation entstehenden Gleichungssysteme mit den verschiedensten Verfahren, wie bei den iterativen Verfahren mit den Überrelaxationsverfahren und bei den direkten Verfahren mit den GAUSSschen Eliminationsmethoden, wird beschrieben. In einem Anhang sind für einige der besprochenen Verfahren FORTRAN-Programme gegeben, einschließlich Hinweise und Erfahrungen bei ihrer Anwendung. — Das Buch gibt eine ausgezeichnete Darstellung der für die numerische Behandlung von partiellen Differentialgleichungen zu entwickelnden Methoden und kann allen an diesen Fragestellungen interessierten Lesern sehr empfohlen werden.

Berlin F. GRUND

**Feldmann, H.**, Einführung in ALGOL 68. Skriptum für Hörer aller Fachrichtungen ab 1. Semester. Braunschweig, Vieweg 1978. IX, 302 S., DM 29,80.

ALGOL 68 ist eine Programmiersprache, die von der bitweisen Informationsverarbeitung bis zur Steuerung paralleler Prozesse umfangreiche algorithmische Möglichkeiten in einem einheitlichen Sprachkonzept bietet. Das vorliegende Buch erläutert die Konstruktion der Sprache an Hand ausführlicher Schemata zur Syntax und vieler Beispiele. Dabei werden Grundkenntnisse auf dem Gebiet der Programmiersprachen vorausgesetzt. Ergänzende Testaufgaben dienen zur Vertiefung und Kontrolle des Verständnisses. Ein Anhang enthält eine Einführung in zweischichtige Grammatiken und die vollständige Syntax von ALGOL 68 im englischsprachigen Original und in deutscher Übersetzung. Damit kann sich der interessierte Leser die stark formalisierte Sprachdefinition aus dem „Revidierten Bericht über die algorithmische Sprache ALGOL 68" (Hrsg. A. VON WIJNGAARDEN u. a. m.) erschließen. Das Buch ist ein sehr wertvoller Beitrag zur weiteren Verbreitung von ALGOL 68.

Berlin M. HARTWIG

**Schäffer, K.-A.** (Hrsg.), Splinefunktionen in der Statistik. Sonderheft zum Allgemeinen Statistischen Archiv 14. Göttingen, Vandenhoeck & Ruprecht 1978. 103 S., DM 19.80.

In vielen Anwendungsbereichen der Technik, der Ökonomie und der Sozialwissenschaften werden Datenmengen aus Messungen abhängiger Größen dahingehend untersucht, ob der Punkteschwarm einer Gesetzmäßigkeit (z. B. Trend, stochastische Funktion etc.) unterliegt. Häufig genügt die Darstellung durch ein Interpolationspolynom. In anderen Fragen werden komplizierte Funktionen nach Auswahl von Stützstellen durch solche einfachen Polynome approximiert. Eine spezielle Klasse von Interpolationsfunktionen mit günstigen Eigenschaften („Glattheit" der Kurve) bilden die spline-Funktionen, deren Gewinnung und Anwendung in verschiedenen Bereichen wie bei Sterbetafeln, bei monatlichen Haushaltseinkommen, bei der Schätzung von Spektraldichten von Zeitreihen sowie von Lag-Verteilungen demonstriert wird. Das Buch ist als Einführung für breite Interessentenkreise, als Spezialliteratur mehr für Ökonomen geeignet.

Berlin H. TOUTENBURG

**de Boor, C.**, A Practical Guide to Splines. Applied Mathematical Sciences 27. Berlin-Heidelberg-New York, Springer-Verlag 1978. XXIV, 392 S., DM 32,50. US $ 17.90.

Hier handelt es sich durchaus nicht um ein „Kochbuch" für den Umgang mit Splinefunktionen. Das Buch setzt einen aufmerksamen, mitarbeitenden Leser voraus. Diesen macht es in sehr persönlicher, interessanter, oft vom Zahlenbeispiel ausgehender Weise mit den Problemen der Polynom-Splines bekannt, regt ihn theoretisch an, ermuntert ihn zu numerischen Experimenten, versichert ihn mit rund 50 (ausführlich kommentierten) Fortranprogrammen. — Der Inhalt ist kurz folgender: Polynominterpolation, stückweise Polynominterpolation, Spline-Interpolation, B-Splines, Probleme der besten Approximation und der Knotenpunktplazierung, Splines zum Glätten, zur Lösung von gewöhnlichen Differentialgleichungen, zur Approximation von Kurven; Tensorprodukt-Splines. — Zu vielen der angeführten Behauptungen werden Beweise gegeben. Abschätzungen sind in der Maximum- bzw. $L_2$-Norm durchgeführt. Jedem Kapitel sind Aufgaben beigefügt, durch die viel zusätzliches Material vermittelt wird. Diese Aufgaben, die Programme und die ausführliche Diskussion der Schwierigkeiten und Möglichkeiten machen das Buch wertvoll.

Berlin G. STOYAN

**Hengartner, W./Theodorescu, R.**, Einführung in die Monte-Carlo-Methode. München-Wien. Carl Hanser Verlag 1978. 188 S., DM 42,—.

Bei der mathematischen Behandlung von immer häufiger auftretenden komplexen Aufgaben, insbesondere in der Physik, Technik und Ökonomie, treten Probleme auf, die sich mit Hilfe klassischer Methoden nicht mehr oder nur mit großem Aufwand lösen lassen. Oftmals kann dann mit Erfolg die Monte-Carlo-Methode eingesetzt werden. Das vorliegende Buch gibt eine Einführung in die Grundlagen und Anwendungsmöglichkeiten dieser Methode. Der Schwerpunkt liegt dabei auf der praktischen Verwendbarkeit der Monte-Carlo-Methode, so daß dieses Buch in guter Ergänzung zur Monographie von ERMAKOW (Die Monte-Carlo-Methode und verwandte Fragen), in der ausführlich die wahrscheinlichkeitstheoretischen Grundlagen der Monte-Carlo-Methode behandelt werden, eine Lücke auf diesem Gebiet, wo es an übersichtlichen und zusammenfassenden Arbeiten mangelt, ausfüllt.

Das Buch unterteilt sich in vier Kapitel. Das erste Kapitel behandelt kurz die Grundlagen der Monte-Carlo-Methode, wobei auch auf Genauigkeit und Wirksamkeit eingegangen wird. Das zweite Kapitel beschäftigt sich ausführlich mit der Erzeugung von Zufallszahlen. Betrachtet werden verschiedene Erzeugungsverfahren für gleichverteilte Zufallszahlen, Transformationsmöglichkeiten, um daraus beliebig verteilte Zufallszahlen zu gewinnen, und damit zusammenhängende Tests. Im dritten Kapitel wird auf die Verwendung der Monte-Carlo-Methode bei Problemen deterministischer Natur eingegangen. Angeführt sind die Berechnung von bestimmten Integralen und Summen, die Lösung linearer Gleichungssysteme, die Lösung von Randwertproblemen und Extremalprobleme. Das vierte Kapitel ist der Behandlung von wirtschaftlich interessanten und praktisch wichtigen Problemen stochastischer Natur gewidmet. Untersucht werden Warteschlangenprobleme, die Simulation der Landung, eines Flugzeugs, Simulationen in der Erdölindustrie, um unbekannte Öl- oder Gasvorkommen aufzufinden, Lernmodellprobleme und Simulationen in der Verschmutzungskontrolle. Bei den behandelten Beispielen wird die Methode soweit aufbereitet (z. B. in Form von Algorithmen und Flußdiagrammen), daß eine weiterführende Nutzung möglich erscheint. In allen Kapiteln werden Hinweise auf mehr ins Spezielle gehende Veröffentlichungen gegeben. Dem Buch, das gut verständlich abgefaßt ist und neben einer allgemeinen Einführung vor allem die vielen Anwendungsmöglichkeiten der Monte-Carlo-Methode aufzeigt, ist ein weiter Leserkreis zu wünschen.

Erkner H. FRIEDRICH

**Wets, R.,** Grundlagen konvexer Optimierung. Lecture Notes in Economics and Mathematical Systems 137. Berlin-Heidelberg-New York, Springer-Verlag 1976. VI, 146 S., 10 Abb., DM 18,–. US $ 8.00.

In dem vorliegenden Buch wird ein ausgewogener Überblick über die konvexe Optimierung gegeben, der dargebotene Stoff entspricht im Umfang einer einsemestrigen Vorlesungsreihe des Verfassers. In Begriffsbildung und Darstellungsweise findet man Parallelen zu R. T. ROCKAFELLAR, Convex Analysis (Princeton University Press, Princeton 1970), jedoch gelingt es dem Verfasser in vorzüglicher Weise, sich in diesem weitverzweigten Feld auf wesentliche Fakten zu beschränken.

In den ersten Kapiteln (I Konvexe Mengen und konvexe Funktionen, II Stetigkeit, III Trennungssätze, IV Konjugierte Funktionen) werden die benötigten Grundlagen der konvexen Analysis bereitgestellt. Ein Teil der Behauptung steht ohne Beweis. Anfangs werden allgemeine reelle Vektorräume $E$, $F$ zugrunde gelegt, später wird spezialisiert zu $E = R^n$, $F = R^m$. Da von vornherein der gesamte Vektorraum als Definitionsbereich der konvexen Funktionen angesehen wird, erfolgt einführend eine Erweiterung der Definition der Addition, so daß brisanterweise u. a. $\infty + (-\infty) = (-\infty) + \infty = \infty$ definiert ist, was z.B. bei ROCKAFELLAR vermieden wird. — Wo wird diese (zusammen mit der Multiplikation nicht widerspruchsfreie) Definition benötigt?

Kap. V (Darstellungssätze) ist Extremalpunkten und Extremalstrahlen konvexer Mengen gewidmet, für den Spezialfall konvexer Polyeder („polyedrische Konvexität") werden die Sätze von MINKOWSKI/WEYL und FARKAS bewiesen. In Kap. VI folgen Alternativsätze für lineare und für konvexe Systeme, u. a. erscheinen Dualitätssätze für lineare Programme sowie Aussagen über Zwei-Personen-Nullsummenspiele. Die Äquivalenz der Sätze von HELLY und CARATHEODORY wird diskutiert. In den nachfolgenden Kapiteln VII und VIII erfolgt die Bereitstellung verschiedener üblicher Differenzierbarkeitsbegriffe (FRÉCHET, GÂTEAUX), die Diskussion von Punkt-Menge-Abbildungen („Multifunktionen") sowie die Untersuchung von Schnitten und Projektionen konvexer Mengen. Das einseitige GÂTEAUX-Differential bezeichnet Vf. als „Hemidifferential".

Mit Kap. IX beginnt die eigentliche Behandlung konvexer Optimierungsprobleme. Es werden notwendige und hinreichende Optimalitätsbedingungen angegeben für (nicht notwendig konvexe) Probleme ohne Restriktionen, die auf Probleme mit Restriktionen übertragen werden mit Hilfe der Indikatorfunktion des zugehörigen Restriktionsbereiches. Die hergeleiteten Optimalitätskriterien haben zunächst hauptsächlich geometrischen Charakter, sie stützen sich wesentlich auf Hemidifferenzierbarkeitsvoraussetzungen. Dieses Konzept erfordert die Einführung der zusätzlichen (nicht gravierenden) Bedingung der „Radialregelmäßigkeit" des Restriktionsbereiches, um u. a. die Hemidifferenzierbarkeit der zugehörigen Indikatorfunktion zu sichern. Ausführlich werden Regularitätsbedingungen diskutiert. Geraffte Darstellungen der LAGRANGE-Theorie und speziell der KUHN-TUCKER-Theorie schließen sich an. Dualitätsuntersuchungen erfolgen ausschließlich für konvexe Optimierungsaufgaben (Kap. X). Eine vorangestellte geometrische Motivation des Dualproblems erleichtert das Verständnis. Neben den üblichen (schwachen und starken) Dualitätsaussagen werden Zusammenhänge zur Störungstheorie und zu Stabilitätsproblemen dargelegt. Das abschließende Kap. XI ist Lösungsverfahren für konvexe Optimierungsaufgaben gewidmet. Die Ausführungen dazu sind knapp, es werden keine einsatzbereiten Algorithmen vorgestellt, sondern man findet „methodische Zugänge" für einzelne Verfahrensklassen (Methoden der zulässigen Richtungen, Approximationsmethoden und Penalty-Methoden). In einem kurzen Anhang gibt Vf. zu den einzelnen Kapiteln einige Quellenhinweise und führt Vergleiche zur Darstellungsweise anderer Autoren. Ein eigentliches Literaturverzeichnis fehlt, ebenso ein Sachwortregister.

Als Leser verspürt man geringfügig, daß der Verfasser Deutsch nicht als Muttersprache beherrscht, was jedoch die Verständlichkeit nicht beeinträchtigt. (Ebenso sind Formulierungsungenauigkeiten einzuordnen: Die Behauptungen 5(i) und 5(iii) in Kap. I sind in der angegebenen Formulierung nicht gleichwertig!) Einzelne Begriffsbildungen weichen von der geläufigen Terminologie ab, z.B. wird mit „Innere Annäherung zu $S$ in $x$" (S. 67) die Menge der zulässigen Richtungen im Punkt $x \in S$ bezüglich der Menge $S$ bezeichnet, der Begriff „Dualitätsbresche" (S. 110) steht für Dualitätslücke u. a.

Das Buch wendet sich vor allem an Mathematiker. Es werden jedoch nur Grundkenntnisse in linearer Algebra und reeller Analysis vorausgesetzt, so daß es evtl. auch Interessierten anderer Fachrichtungen zugänglich ist. Die Darstellung ist didaktisch gelungen und trotz ihrer konzentrierten Form gut lesbar. Es liegt eine exakte, geraffte, prägnante und fundierte Einführung in die Optimierungstheorie vor, die nicht nur (entgegen der Ankündigung im Titel) den konvexen Fall erfaßt. Das Buch vermittelt viele Anregungen. Es kann jedem Interessierten wärmstens empfohlen werden.

Magdeburg F. JUHNKE

**Henn, R. / Korte, B. / Oettli, W.** (Hrsg.), Optimization and Operations Research. Proceedings of a Workshop Held at the University of Bonn, Oct. 2–8, 1977. Lecture Notes in Economics and Mathematical Systems 157. Berlin-Heidelberg-New York, Springer-Verlag 1978. VI, 270 S., DM 27,–. US $ 13.50. ISBN 3-540-08842-3

This volume presents 27 lectures given at a workshop organized by the Institute of Operations Research at the University of Bonn (FRG). The lectures, all being at a high level, inform the reader very well on the state of the art (with respect to 1977) and on new results in the respective fields and doubtless are of importance still today. So the book can be recommended to the interest of researchers working in the field of Operations Research and especially in Mathematical Programming. — It is not possible here to list all the titles of the lectures, but to group them around some central subjects: A good part of the papers is devoted to integer programming, matroids and related topics like networks and graphs (A. BACHEM, S. BAUM and L. E. TROTTER JR., P. BRUCKER, R. E. BURKARD and U. ZIMMERMANN, U. DERIGS, R. EULER, M. GRÖTSCHEL and M. W. PADBERG, R. KANNAN and C. L. MONMA, L. MATTHEWS, C. P. SCHNORR). Another part of lectures treats optimal control, operator equations etc. (L. COLLATZ, J. GWINNER, PH. MICHEL, E. J. MUTH and K. SPREMANN, E. SACHS, M. VLACH) as well as nonlinear programming (J. HARTUNG, D. HAUSMANN and B. KORTE, P. KOSMOL, F. LEMPIO and H. MAURER, J. ZOWE and H. MAURER, and, in a certain sense B. DEJON), where overlapping cannot be avoided or is natural. The rest of the paper has more or less marginal character in the context of this workshop and concerns parametric linear programming (B. BROSOWSKI), several stochastic aspects (D. BIERLEIN, W. GAUL), especially inventory (K. HINDERER and D. KADELKA) and queueing problems (M. SCHÄL).

Berlin H. WEINERT

**Kushner, H. J. / Clark, D. S.,** Stochastic Approximation Methods for Constrained and Unconstrained Systems. (Applied Mathematical Sciences 26). Berlin-Heidelberg-New York, Springer-Verlag 1978. X, 261 S., 4 Abb., DM 26,40. US $ 13.20.

Die stochastische Approximation findet seit dem Erscheinen der grundlegenden Arbeiten von ROBBINS-MOURO („A stochastic approximation method", Ann. Math. Statist. **22** (1951)) und KIEFER-WOLFOWITZ („Stochastic estimation of the maximum of a regression function", Ann. Math. Statist. **23** (1952)) sowohl im Hinblick auf ihre Anwendbarkeit als auch bezüglich theoretischer Überlegungen lebhaftes Interesse. H. J. KUSHNER und D. S. CLARK — ersterer durch zahlreiche Publikationen hochqualifiziert für ein solches Vorhaben — stellen mit dem vorliegenden Buch der großen Reihe von Publikationen und Monographien (z. B. von FABIAN, SCHMETTERER, WASAN, NEVELSON-CHASMINSKIJ) einen sehr spezifischen Beitrag zur Seite. Diese Spezifik besteht einerseits in dem Versuch einer recht einheitlichen Beweisführung bzgl. der Konvergenz des ROBBINS-MOURO-Prozesses (RMP) sowie des KIEFER-WOLFOWITZ-Prozesses (KWP), die verhältnismäßig einfache Kompaktheitsmethoden einschließt. Damit ist andererseits das Bemühen verbunden, die i. a. hinreichenden Konvergenzbedingungen weiter zu „minimieren". Dies führt hier zu einer beträchtlichen Erweiterung der bisher bekannten Resultate. Im einzelnen behandeln die Autoren nach einer Einleitung (I.) (RMP — eine stetige Version des RMP — Regelung eines dynamischen Systems (Bsp.) —

Funktionsminimierung (KWP — Probleme mit Nebenbedingungen)) die folgenden Themen:

II. Konvergenz mit Wahrscheinlichkeit Eins für Systeme ohne Nebenbedingungen (RMP und KWP: Bedingungen — Konvergenzbeweise für RM- und KW-strukturierte Algorithmen — ein allgemeiner RMP („exogenes" und zustandsabhängiges Rauschen) — Anwendungen — Menšov-Rademacher-Schätzungen).

III. Schwache Konvergenz von Wahrscheinlichkeitsmaßen.

IV. Schwache Konvergenz für Systeme ohne Nebenbedingungen (RMP — KWP — ein allg. RMP („exogenes" und zustandsabhängiges Rauschen) — Identifikationsproblem — Dichtheit („Tightness")).

V. Konvergenz mit Wahrscheinlichkeit Eins für Systeme mit Nebenbedingungen (Strafmethoden-Algorithmus („Penalty-Multiplier") bei Nebenbedingungen in Gleichungsform — Lagrangesche Methode bei Nebenbedingungen in Ungleichungsform — Projektionsalgorithmus — Strafmethode für Nebenbedingungen in Ungleichungsform).

VI. Schwache Konvergenz: Systeme mit Nebenbedingungen (ein Algorithmus vom Multiplikatortyp für Nebenbedingungen in Gleichungsform — Lagrangesche Methode — Projektionsmethode — Strafmethode für Nebenbedingungen in Ungleichungsform).

VII. Konvergenzgeschwindigkeiten (KWP, RMP).

Obwohl einige Anwendungsbeispiele benannt werden, handelt es sich hier um eine Monographie, die sich eindeutig den theoretischen Aspekten der stochastischen Approximation verpflichtet fühlt. Beim Leser werden fundierte Kenntnisse auf dem Gebiet der Maß- und Wahrscheinlichkeitstheorie vorausgesetzt. Die Darstellung ist außerordentlich konzentriert gehalten; die Lektüre dürfte also nur bei sorgfältigem Durcharbeiten von Gewinn sein. Die vorliegende Monographie ist folglich in erster Linie für Leser geeignet — und zu empfehlen —, die selbst auf dem Gebiet der stoachastischen Approximation arbeiten oder die Entwicklung dieser Theorie mit ständigem Interesse verfolgen.

Dresden V. Nollau

**Neuts, M. F.** (Editor), Algorithmic Methods in Probability. Tims Studies in the Management Sciences 7. Amsterdam-New York-Oxford, North-Holland 1977. VI, 296 S., US $ 29.00. Dfl. 65.00.

Den 17 Arbeiten des gut ausgestatteten Sammelbandes ist gemeinsam, daß in ihnen stochastische Probleme numerisch oder — über Abschätzungen — bis zu auswertbaren Formeln gelöst werden. Die verwendeten Methoden variieren dabei entsprechend den Aufgabenstellungen sehr stark. So gibt es Arbeiten über numerische Umkehrung von Funktionaltransformationen, numerische Optimierung bei Maximum-Likelihood-Schätzungen, Auswahl optimaler Systeme mittels Simulation und numerische Behandlung von Bedienungssystemen, die durch eingebettete Markowsche Ketten beschreibbar sind. Natürlich ist die breite Anwendung numerischer Verfahren in der angewandten Wahrscheinlichkeitsrechnung eine Notwendigkeit; dennoch erscheint mir bei einigen Arbeiten des Sammelbandes der numerisch-rechentechnische Aufwand angesichts der Praxisferne der untersuchten Modelle als zu hoch. Liste der Autoren und Arbeiten:

M. F. Neuts, Preface; H. Bohman, Numerical Fourier Inversion; B. Allen; R. S. Anderssen; E. Seneta, Computation of Stationary Measures for Infinite Markov Chains; D. Hunter, Approximating Percentage Points of Statistics Expressible as Maxima; D. L. Iglehart, Simulating Stable Stochastic Systems, VII: Selecting the Best System; E. J. Dudewicz; N. A. Zaino, Jr., Allowance for Correlation in Setting Simulation Run-Length via Ranking-and-Selection Procedures; S. H. Zanakis, Computational Experience with Some Nonlinear Optimization Algorithms in Deriving Maximum Likelihood Estimates for the Three-Parameter Weibull Distribution; H. Moskowitz; R. K. Fink, A Bayesian Algorithm Incorporating Inspector Errors for Quality Control and Auditing; R. J. Kryscio; N. C. Severo, Statistical Inferences for a Stochastic Epidemic Model Proposed by Gart; M. Reiser, Numerical Methods in Separable Queueing Networks; E. J. Muth, Numerical Methods Applicable to a Production Line with Stochastic Servers; B. Pagurek; C. M. Woodside, A Recursive Algorithm for Computing Serial Correlations of Times in an M/G/1 Queue; M. F. Neuts, Algorithms for the Waiting Time Distributions under Various Queue Disciplines in the M/G/1 Queue with Service Time Distributions of Phase Type; O. S. Yu, The Steady-State Solution of a Heterogenous-Server Queue with Erlang Service Times; D. M. Avis, Computing Waiting Times in GI/$E_k$/c Queueing Systems; F. A. Al-Khayyal; D. Gross, On Approximating and Bounding GI/M/c Queues; S. N. Raju; U. Bhat, Recursive Relations in the Computation of the Equilibrium Results of Finite Queues; J. E. Knepley; M. J. Fischer, A Numerical Solution for Some Computational Problems Occurring in Queueing Theory.

Freiberg D. Stoyan

In der DDR beziehbar im Rahmen des Kontingents.

# EINGEGANGENE BÜCHER

Bei der Schriftleitung sind folgende Bücher eingegangen (ausführliche Besprechung bleibt vorbehalten).

**Bibikov, Y. N.,** Local Theory of Nonlinear Analytic Ordinary Differential Equations. Lecture Notes in Mathematics 702. Berlin-Heidelberg-New York, Springer-Verlag 1979. IX, 147 S., DM 21,50, US $ 11.90. ISBN 3-540-09114-9

**Postlethwaite, I. / MacFarlane, A. G. J.,** A Complex Variable Approach to the Analysis of Linear Multivariable Feedback Systems. Lecture Notes in Control and Information Sciences 12. Berlin-Heidelberg-New York, Springer-Verlag 1979. IV, 177 S., 53 Abb., DM 21,50, US $ 11.90. ISBN 3-540-09340-0

**Bergh, J. / Löfström, J.,** Interpolation Spaces. An Introduction. Grundlehren der mathematischen Wissenschaften 223. Berlin-Heidelberg-New York, Springer-Verlag 1976, X, 207 S., 5 Abb., DM 60,—, US $ 24.60. ISBN 3-540-07875-4

**Maravall, A.,** Identification in Dynamic Shock-Error-Models. Lecture Notes in Economics and Mathematical Systems 165. Berlin-Heidelberg-New York, Springer-Verlag 1979. VIII, 158 S., DM 21,50, US $ 11.90. ISBN 3-540-09112-2

**Sawaragi, Y. / Soeda, T. / Omatu, S.,** Modelling, Estimation, and Their Applications for Distributed Parameter Systems. Lecture Notes in Control and Information Sciences 11. Berlin-Heidelberg-New York, Springer-Verlag 1978. VI, 269 S., 4 Abb., DM 28,50, US $ 15.70. ISBN 3-540-09142-4

**Hwang, Ching-Lai / Masud, A. S. Md.,** Multiple Objective Decision Making — Methods and Applications. A State-of-the-Art Survey. In Collaboration with S.R. Paidy and Kwangsun Yoon. Lecture Notes in Economics and Mathematical Systems 164. Berlin-Heidelberg-New York, Springer-Verlag 1979. XII, 351 S., 39 Abb., 35 Tab., DM 35,50, US $ 19.60. ISBN 3-540-09111-4

www.ingramcontent.com/pod-product-compliance
Lightning Source LLC
LaVergne TN
LVHW081350011026
849161LV00013BB/3041
* 9 7 8 3 1 1 2 6 4 9 9 5 4 *